P9-BVG-939

FIRE, WATER, AND AIR

the story of

Antoine Lavoisier

FIRE, WATER, AND AIR

the story of

Antoine Lavoisier

Roberta Baxter

MORGAN REYNOLDS

PUBLISHING

Greensboro, North Carolina

the

Profiles
IN SCIENCE

series includes biographies about . . .

Nikola Tesla
Louis Pasteur
Tycho Brahe
Johannes Kepler
Nicholas Copernicus
Galileo Galilei
Isaac Newton
Robert Boyle
Rosalind Franklin

Ibn al-Haytham
Edmond Halley
Marie Curie
Caroline Herschel
Thomas Edison
Michael Faraday
Antoine Lavoisier
Charles Darwin

FIRE, WATER, AND AIR
THE STORY OF ANTOINE LAVOISIER

Library of Congress Cataloging-in-Publication Data
Baxter, Roberta.
 Fire, water, and air : the story of Antoine Lavoisier / by Roberta Baxter.
  p. cm. -- (Profiles in science)
 Includes bibliographical references and index.
 ISBN-13: 978-1-59935-087-5 (alk. paper)
 ISBN-10: 1-59935-087-4 (alk. paper)
 1. Lavoisier, Antoine Laurent, 1743-1794. 2. Chemists--France--Biography.
3. Chemistry--France--History--18th century. 4. Chemistry--Nomenclature. 5.
Chemical processes. I. Title.
 QD22.L4B24 2008
 540.92--dc22
 [B]
                              2008041219

Printed in the United States of America
First Edition

Contents

Antoine Lavoisier
(Courtesy of North Wind Picture Archives/Alamy)

one

Early Life

When Antoine Lavoisier was starting his scientific studies, most scientists believed that there were four elements: water, fire, air, and earth. The elements could be turned into each other if the right process was found. In 1661, English chemist Robert Boyle had shown this theory to be false, but many still believed it. Boyle disproved the four element theory, but he did not have a unified theory to take its place, so the ancient ideas continued. Scientists conducted experiments to turn air into water or earth into fire.

One example seemed to prove that water turned into earth. When water was heated for a long time, solid material (a presumed element of earth) formed. At age twenty-five, Lavoisier decided to test this experiment.

He began with rainwater, the purest water available. To increase the purity, he distilled it [boiled it and captured the vapor] eight times. For the next part of the experiment, he

Famed chemist Robert Boyle disproved the four element theory. *(Courtesy of Granger)*

chose a vessel known as a pelican. The glass jar has two arms and a stopper. As the water heats up, the vapor rises into the top and the arms, condenses, and runs back down.

One of the most important instruments in Lavoisier's laboratory was a balance. He believed he could only understand the results of his experiments by precise measurements of the weight of the substances. Lavoisier meticulously weighed the pelican and the water that he put into it.

He carefully heated the vessel with the stopper loose to allow the expanding air to escape to prevent the pelican from breaking. When he figured that the air had expanded as much as possible, he sealed the pelican and continued the heating. The pelican was placed in a sand bath heated by an oil lamp, providing a temperature that kept the water almost boiling. The experiment started on October 24, 1768. For 101 days, Lavoisier continued heating the water.

For the first month, no change was seen. But on November 20, Lavoisier saw some thin flakes floating in the water. The particles were so small they could only be seen with a magnifying glass. Lavoisier kept the heat on. After December 15, the flakes migrated to the bottom of the vessel and stayed there.

On February 1, 1769, Lavoisier turned off the lamp and allowed the pelican to cool. When the pelican with the water inside was weighed, it was the same as the combined weight of the pelican and the water at the start of the experiment.

When Lavoisier removed the stopper, he heard a rushing of air moving into the pelican, telling him that the stopper had held during the experiment. He carefully poured the water and the flakes into other vessels, dried the pelican, and weighed it. It had lost weight. Then he weighed the water and found that it was the same weight. Finally, he weighed the tiny flakes. Their weight was close to the amount of weight that the pelican had lost. Lavoisier realized that the water had not turned into "earth." The flakes had come from the glass of the pelican. He reported to the Academy of Science "that the nature of water is not changed, nor does water acquire any new properties through repeated distillation."

Just as Lavoisier used his balance to determine that water did not change into earth, he looked for balance or equilibrium in other areas of his life. He managed to balance work in his laboratory with service to the government. He advocated balance in the fields of economics, farming, and education. Yet the unbalance of the political world in which he lived led to the French Revolution. That time of chaos led to the end of Lavoisier's experiments and contributions to the scientific world.

Antoine Lavoisier was born into a family that had raised itself from a peasant background. The family originally lived in Villers-Cotterêts, France, about fifty miles northeast of Paris. An early ancestor had cared for the king's horses. Another was a postmaster, and a third was a sheriff. The sheriff's son, Nicolas, became a merchant and pulled the family into the middle class. His son, Antoine, became a lawyer and so did his son, Jean Antoine. (For six generations before the chemist Antoine, there had been a son named Antoine Lavoisier.)

Antoine's father, Jean Antoine Lavoisier, became a lawyer and inherited his uncle's position as an attorney to the Parliament of Paris. He also inherited his uncle's house in Paris. Jean Antoine married Èmilie Punctis, the daughter of a wealthy Parliament lawyer.

In those days, people's position in society depended on class. The people of France were divided into three classes. Each class was known as an Estate. The First Estate was the clergy or church officials. Many of these men were wealthy. The Second Estate was the nobles. These men were wealthy landowners who owned property that had been passed down in their families. The nobles and the king controlled most of the money in France. The Third Estate was everyone else, from prosperous merchants, lawyers, and shop owners, to the poor peasants.

The Lavoisier family belonged to the Third Estate in the middle-class section. Their wealth had been growing through the years, but they had not reached the level of nobility.

Antoine was born on August 26, 1743, in Paris. His sister, Marie, was born two years later. Not much later though, his

A drawing depicting the Three Estates of French Society: clergy, nobles, and commoners *(Library of Congress)*

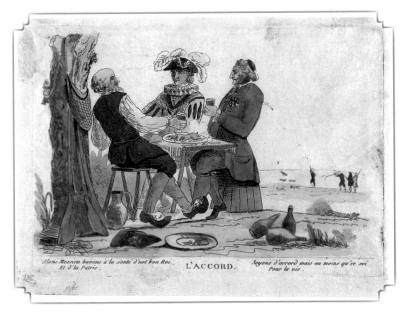

maternal grandfather died; and when Antoine was five years old, his mother died.

The family pulled inward in their grief. Antoine's father moved his family into his mother-in-law's house. Also living with Grandmother Punctis was her unmarried daughter, Marie Marguerite Constance. Antoine's grandmother and aunt cared for him and his sister. Although Antoine's father was only in his thirties when his wife died, he never remarried. He and his son formed a close bond; it was strengthened further in 1760, when Antoine's sister Marie died at the age of fifteen.

Lavoisier studied at College de Quartes-Nations, also known as College Marzarin for the founder of the school. Giulio Marzarin was an Italian, who became a French naturalized citizen. He offered his services to the French king and in 1642, became prime minister. After the death of King Louis XIII, the prince, Louis XIV, became king. Because the prince was only a child, Marzarin governed the country until Louis XIV grew old enough to reign. When Marzarin died, he left money in his will to start the College de Quartes-Nations.

At the school, studies included languages, literature and composition, and science. Antoine received awards for Greek, Latin, and French composition. In 1760, Antoine received second place in an essay competition among all French public schools. Unfortunately no copies of the essay entitled "Whether rectitude of the heart is as necessary as precision of intelligence in the search for truth" still exist.

While at College Marzarin, Antoine attended science lectures during his senior year. One teacher, Nicolas Louis de Lacaille taught mathematics and tutored some students in astronomy. One of Lacaille's achievements was measuring

Lavoisier studied at the College de Quartes-Nations.

a distance along the meridian, the line running north and south on the earth. Using that measurement, he calculated the circumference of the earth. Antoine studied the stars with Lacaille.

Despite a burgeoning interest in the sciences, after finishing at Marzarin, Antoine decided to follow in the footsteps of his father and grandfathers and study law. He entered the University of Paris and received a bachelor of law in 1763.

While studying law, Antoine continued attending science lectures. Lectures were given at Jardin du Roi (the King's Garden), and they were open to the public. First, a science professor would discuss a topic, and then a demonstrator would perform experiments. Professors turned up their noses at getting their hands dirty with experiments, but the demonstrators were the favorite part of the lecture for most people. When Antoine attended the chemistry lectures, the demonstrator was Guillaume Rouelle.

Rouelle's demonstrations often proved the opposite of what the professor had taught at the beginning of the lecture. Rouelle was a flamboyant person. When starting an experiment, he would be properly dressed and calm. As he worked, his excitement would grow and off would come his hat and then his wig, both usually hung on some equipment on the

A 1636 painting of the Jardin du Roi, where Lavoisier attended lectures.

An 1874 engraving of Rouelle making a chemical explosion during a chemistry demonstration. *(Courtesy of The Print Collector/Alamy)*

stage. Eventually, as he worked, he would fling off his coat and vest and continue his experiments in his shirt. Many times, Rouelle would mention his own ideas saying, "That is one of my secrets that I tell to nobody."

The lectures were attended by people from around Paris, and Rouelle's enthusiasm ignited interest in chemistry in many. Antoine was lucky enough to be able to get a copy of notes taken during Rouelle's lectures, so he had a permanent copy of the man's ideas and experiments.

Antoine became interested in meteorology, the study of weather. He began to take several barometric readings a day and even persuaded his family to keep up the record when he was away. Years later, he persuaded people across France and even as far as Baghdad to keep barometric records. He concluded that a person might be able to predict the weather with enough barometric readings and wrote a set of rules for the forecast. He added that a person would also need to know the strength and direction of the wind and the amount of moisture in the air. Today, meteorologists use exactly these measurements to forecast the weather.

In 1763, while still a law student, Antoine became an assistant to Jean Ètienne Guettard, a famed geologist. The two would tramp out into the countryside and collect samples of soil, rock, and water. Guettard was interested in the formations they found, but also in the chemical components of the rocks and minerals. Antoine went on the expeditions when he had time between his law studies.

Law, however, was not to be Antoine's work. The study of science would be his life, Antoine decided. He intended to do his best to understand the science that had intrigued him since his attendance at Rouelle's lectures. He stated, "I am young and avid for glory."

two

Early Science

O nce he had chosen science as his life's work, Antoine Lavoisier desired to be the best scientist he could, and be accepted by other scientists. The most prestigious scientific group in France was the Academy of Sciences. To be chosen to join the Academy was the highest honor a scientist in France could have.

The Academy of Sciences was made up of forty-two members, divided into six scientific sections. Each section had seven people, and new members could only be added when a vacancy occurred because of death or a junior person moving into a more advanced position. A scientist was nominated to fill a vacancy by an election of the members and then approved by the king.

Most of the members were much older than Lavoisier, but he was determined to show them what he could do.

His campaign for membership began with two scientific studies.

Lavoisier studied the mineral gypsum. When gypsum is heated, it becomes a powder. When that powder is mixed with water, it hardens into plaster of Paris. Lavoisier's interest in gypsum was aroused during his studies with the geologist, Jean Ètienne Guettard.

The substance had been used for centuries by people including ancient Egyptians and Greeks. It later got the name "plaster of Paris" because of abundant supplies of gypsum around Paris. After a great fire destroyed large sections of London, England, in 1666, the king of France declared that all wooden houses in Paris should be coated with plaster of Paris to make them more fireproof. People knew how to turn gypsum into plaster of Paris, but not what occurred in the reaction.

Lavoisier analyzed the solubility of gypsum from different places and rock layers. By careful weighing, he determined that gypsum lost weight when it was heated. He concluded that it was water that was driven off, released from the mineral which then disintegrated into powder. The amount of the water lost was about one-fourth of the original weight of the mineral. When water was added back to the powder, the mixture hardened into plaster of Paris. Lavoisier stated that the water and the crystals that formed were the cause of the solidification of the gypsum.

Lavoisier noted that if gypsum was overheated, it would not form plaster of Paris when water was added. He was not sure why this was true, but he presented it as a challenge to understand. He believed that he should not guess, but only present what he could learn from experiments.

Rouelle, the demonstrator who had ignited Lavoisier's interest in science, had studied crystals. Lavoisier acknowledged that his own investigation was built on the work of men like Rouelle. He wrote, "There is a certain sequence in human knowledge that cannot be broken and that is crucial to the success of our discoveries."

Lavoisier presented two papers he wrote about gypsum to the Academy of Sciences in February 1765 and April 1766. Scientists would speak about their research and answer questions before the Academy of Sciences. When a person had presented several papers, he was considered enough of a scientist to possibly be a member of the Academy.

Lavoisier, dressed in his best clothes and a powdered wig, performed his experiments on gypsum before the Academy. The older, more experienced scientists watched carefully. At the end, they agreed that his presentation was well done and presented new information. Lavoisier had impressed the Academy.

In 1764, the government was searching for a way to better light the streets of Paris. The Academy began a competition for the best way. Scientists had a year to study methods and devise a system. Judges of the contest would be members of the Academy of Sciences. A money prize would be given to the scientist who won the competition. Lavoisier took the challenge.

He tested different types of lamps, burned various candles and oils, and devised reflectors to increase the amount of light. He measured how far each type of light would cover. For each test, he calculated the cost. Lavoisier even lived for six weeks in a dark room so he could more easily detect a difference in the amount of light put out by a lamp.

After his experiments, Lavoisier wrote an essay listing his results and conclusions. Included was a list of the methods he tried, the type of oil and shape of lamp he recommended, and the costs of each system. He also covered the history of street lighting in Paris and other cities.

When the essays were judged, the Academy divided them into two types. One type offered practical solutions to street lighting and the other group was mostly theoretical, applying math and physics to the problem. Lavoisier's essay was in the second category.

When the award was given, the money was divided between three essays from the practical group. These three essays were submitted by manufacturers of street lights and the three companies split the award money. However, Lavoisier's essay had impressed the judges enough that they presented him with a special medal from the king. The gold medal was presented to Lavoisier on the stage of the Academy on April 9, 1766. The *Journal des Savants,* a magazine from the Academy, said, "this flattering distinction for so young an author, of which there is no previous instance in the Academy, has greatly pleased the public."

Lavoisier was only twenty-three years old, but he had made an impact on science and the scientists of the Academy. In 1766, a junior position in the chemistry department of the Academy was vacant when a member moved up to be an associate. Lavoisier campaigned for the position and his father and aunt spoke to their friends who had influence. However, when the voting was finished, Lavoisier was not chosen. He was still very young and his chance would come again.

Lavoisier had chosen science as his life's work, not the law. Events in his family made this choice even easier. His grandmother Punctis died. In those days, a person was not considered adult enough to receive an inheritance until age twenty-five. However, Lavoisier's father arranged for him to be emancipated, meaning he could receive the inheritance from his mother and grandmother. With the inheritance at age twenty-three, Lavoisier had enough money to support himself while he pursued a life in science.

During his law studies and early scientific lessons, Lavoisier had assisted Jean Guettard with geologic studies in the Paris area. Now Guettard offered him a thrilling opportunity: the government had agreed to support Guettard's plan for a geological atlas of France. He asked Lavoisier to go with him on a long tour to the Vosges Mountains, far from Paris along the French-Swiss border. They would collect mineral and water

Lavoisier traveled the Vosges Mountains with Jean Guettard.

samples and draw maps of the countryside. Lavoisier accepted with excitement.

The two scientists prepared their belongings and scientific equipment. Because many of the roads they would travel were too rough for carriages, they started off on horseback on June 14, 1767. Lavoisier's father and aunt waved good-bye; they were worried about the dangers of the long trip, but Lavoisier promised to write. He wrote them a letter the first night assuring them that he was well. As the journey continued, Lavoisier was homesick for his family, but intrigued by the scientific work that he and Guettard were doing.

Between five and six o'clock each morning, Lavoisier would record the readings of the thermometer and barometer. During the day, he made observations of the temperature and barometric pressure again and wrote down a final amount in the evening. In his diary and the scientific notebooks, he recorded the soil types and the shape of the land. Lavoisier and Guettard visited quarries and recorded their location for the map that they would draw from the information. They tested the temperature and density of the water from rivers and springs and Lavoisier tested the waters for minerals. The density of water varies with the amount of dissolved minerals. He even performed these tests on the water at the inns where they stayed. Lavoisier wrote to his father, "when I get back to Paris I will have made complete analyses of samples of all the natural and mineral waters of this region."

He regretted that traveling on horseback prevented him from bringing along other scientific instruments. He thought he could have performed other tests if he had more equipment.

Along the way, Lavoisier and Guettard stayed where they could find room. In one village, the only place for them to

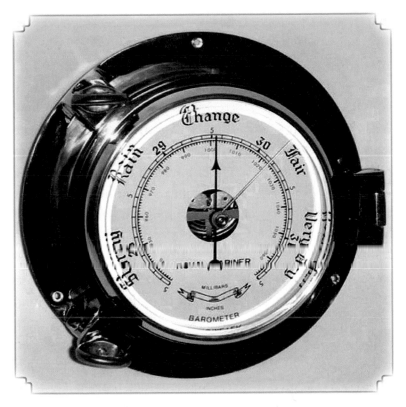

A barometer, similar to the one used by Lavoisier.

sleep was a farmer's storage shed. It gave them a roof over their heads, but unfortunately it was also full of onions. As Lavoisier wrote in a letter, the smell "tainted" them.

The Vosges Mountains thrilled Lavoisier. He wrote, "I have never seen anything in Nature that has impressed me so much." He climbed the 3,800-foot Ballon d'Alsace, one of the highest peaks in the area.

Lavoisier's aunt worried about the dangers he faced. She wrote, "Fear about the mines at Sainte-Marie obsesses me; what a relief for me when you will be on your way back from those mines, which cause all my worries!" She encouraged him to continue sending letters. "A letter scarcely reaches us but we are already waiting for the next." As it happened, the mines were flooded, so Lavoisier didn't enter them.

His father also wanted him to keep sending letters. He wrote "Send news as often as you can, a word in your hand that you are well with the date and place from which you write . . . You know how dear you are to us."

Near the city of Basle, Switzerland, they traveled along the Rhine River. Lavoisier wrote about its swift flowing beauty. In Basle, they met Daniel Bernoulli, a professor at the University of Basle. Bernoulli became a famous mathematician and studied hydrodynamics, the study of how fluids move.

An 1840 painitng of the town of Bingen, Germany, along the Rhine River. Lavoisier and Guettard traveled along the Rhine during their journey.

DANIEL BERNOVLLIVS

Daniel Bernoulli

After climbing the Grand Ballon, Lavoisier and Guettard continued on to Strasbourg, France. Lavoisier met other scientists there, including a man who also collected barometric readings and sent the measurements to Lavoisier by mail.

Lavoisier bought several scientific books in Strasbourg, even though they were written in German. He couldn't buy the same books in Paris, so he figured the expense of the books and shipping them home was worth it.

On this trip through the countryside, Lavoisier saw first-hand how miserable the life of the peasants was. The ancient farming methods provided little food in most years, even with hard work. The social structure kept peasants in poverty, but responsible for taxes that were not paid by the wealthy. He determined to change life for them if he had the chance.

Lavoisier wrote his father that he and Guettard planned to be back to Bourbonne-les-Baines about October 6. His father waited there at the house of friends to reunite with his son.

The expedition of Guettard and Lavoisier had covered more than a thousand miles in four months. They collected an immense amount of information and samples. For several months, Lavoisier was busy organizing the information and writing a report on his tests on water that they encountered. He labeled mineral samples for the government and also for his private collection.

Together Guettard and Lavoisier assembled the information into sixteen sheets of a proposed geologic map of France. The project was expensive and before it was finished, Guettard retired. In 1780, a government official, Monnet, published the incomplete report with thirty-one maps based on information gathered by Guettard and Lavoisier. However, the report only listed Guettard and Monnet, completely ignoring the work that Lavoisier had done. Lavoisier was irritated by this treatment but could do nothing about it.

The expedition did again bring Lavoisier to the attention of the members of the Academy of Sciences. While he was on the expedition, friends had told his father that the members of the Academy were impressed with Lavoisier's scientific studies. His father wrote about a conversation with a friend: "He regrets that you were not admitted to the Academy in the last election and says that since he voted for you, it is not his fault." He added, "I think you have good reason to be hopeful."

In 1768, a member of the Academy, chemist Théodore Baron, died, creating a vacancy. Lavoisier appeared to be a good candidate for the new position. Another chemist, Gabriel Jars, also looked like a strong possibility; Jars was a metallurgist who had supervised opening mines in parts of France and visited mines in other countries, bringing back processes to increase mine production. Lavoisier was only twenty-five, but he had received a medal for his essay on street lighting, written four papers on various topics that had been presented to the Academy, and participated in the geologic expedition.

Lavoisier stayed busy while waiting for the results of the election of the Academy. He built on the results of the expedition in his own research. In March 1768, he gave a report to the Academy about how to measure the density of water at different temperatures. He described the instruments he used in his experiments and his results.

This report indicated the importance of instruments and measurements to Lavoisier. It also pointed out the difficulties of working with liquids. He wrote, "Chemists have plenty of ways of precisely determining the quantities of solid . . . in their experiments. The balance provides reliable tests that cannot be mistaken. But there are problems with certain salts that cannot be reduced to a solid form."

For these substances, Lavoisier realized that weighing them only gave the combined weight of the chemical and the water in which it was dissolved. To know the amount of acid in a sample, "the hydrometer can be wonderfully helpful." A hydrometer measures the density of a liquid. Lavoisier proposed that tables of hydrometer readings be arranged in tables, allowing scientists to determine if the relationship between strength and density was true.

In his report, he also pointed out challenges. "This part of chemistry is much less advanced than is thought; we hardly know the first elements. Every day we combine acids and alkalis, but how do these two substances unite? . . . If it is possible for the human mind to penetrate these mysteries, it can hope to do so through research on the specific weights [densities] of fluids."

(Today it is known that, in general, the stronger the acid, the higher density the liquid mixture has.)

The vote for the vacancy in the Academy of Sciences took place on May 18, 1768. However, the vote was only for the nomination; the final choice was the king's. Both Jars and Lavoisier were nominated, but Jars was ultimately selected for the open position. However, Lavoisier was admitted in a temporary position with the understanding that when a vacancy occurred, he would become a permanent member of the Academy.

The members of the Academy met twice a week, on Wednesdays and Saturdays. Lavoisier attended these meetings and was assigned several projects to report on. Members of the Academy were tasked by the king to study things such as water quality, prisons, fossils, improving mines and factories, and many other topics. Lavoisier enthusiastically joined in. The idea of studying science and helping society fulfilled his greatest dreams.

In 1768, Lavoisier joined another group that would change his life. He invested money in the *Ferme-Générale*. The name means Farmers-General, but most people called it the Tax Farm. Sixty men belonged to this company.

The French government farmed out or contracted to this group the right to collect taxes. Each year, the members paid money to the government for the privilege of collecting taxes and any money they made over that amount and what was owed the government was profit. The system was an investment to the Farmers-General and a way to have taxes collected for the government.

Lavoisier borrowed 500,000 francs (about $95,000) to buy part of a share of the Farmers-General, adding the duties of a Farmer-General to his activities of the Academy of Sciences.

A portrait of Lavoisier on a Tax Farm document.

The government expected the Tax Farm to collect taxes on salt, tobacco, vegetables, grain, fruits, and many other items. Tax collectors spread out over the country to collect taxes, mostly from the poor and middle class. The wealthy were not required to pay taxes.

The Tax Farm employed about 30,000 people to collect taxes. Many of these collectors were unfair. Because they could keep money above the taxes required by the government, some tax collectors extorted extra money from the people. In Paris, a farmer had to pay taxes on vegetables he was bringing into the city to sell. He had to pay the tax before he sold the crop, so if he didn't sell his produce at a high enough price, he lost money.

The unfair methods of tax collecting would determine Lavoisier's life direction in the future, but for the present, he worked hard on several committees to make the system as fair as possible. His connections in the Tax Farm led to another change in his life.

The Tax Farm director, Jacques Alexis Paulze, was impressed by Lavoisier and they became friends. Paulze had a daughter, fourteen-year-old Marie Anne Pierrette Paulze.

Marie was a pretty girl with brown hair and blue eyes. A powerful relative wanted her to marry a friend of his. Marie considered the middle-aged penniless man "a fool . . . a kind of ogre." Her father did not want to force Marie into marriage with such a man, but his relative was pressuring him to make it happen. Paulze saw the unmarried Lavoisier as a solution to the problem. Lavoisier and Marie agreed to marry.

In November 1771, marriage arrangements were made between Lavoisier and Marie Paulze. On December 16, 1771, the fourteen-year-old bride married her twenty-eight-year old groom. They moved into a house bought by Lavoisier's father.

Marie turned out to be a perfect wife for Lavoisier. Many young women her age were running households in Paris. Marie became a good home manager, but she also helped Lavoisier with his scientific work. She took notes and drew pictures to illustrate the experiments, and she learned English so she could translate scientific reports for Lavoisier. She entertained visitors to their home, including scientists from around the world.

With membership in the Academy of Sciences and the Tax Farm, Lavoisier's life direction was set. He continued his scientific experiments and his work on committees for both the Academy and the Tax Farm.

three
Water, Fire, and Air

As he was starting his scientific career, Lavoisier performed the experiment of heating water for 101 days. Lavoisier would perform more experiments with water later. But fire and heat were the focus of his next experiments. Always, he counted on experiments to reveal the truth. Years later, in his most famous book, *Elements of Chemistry,* he wrote:

> Reason must continually be subjected to experimental proof. We must preserve only those facts that are given by nature, which cannot deceive us. Truth must only be sought in the natural connection between experiments and observations, in the same way that mathematicians arrive at the solutions of a problem by a simple arrangement of the givens. By reducing reason to the simplest possible operations and restricting judgment as much as possible, they avoid losing sight of the evidence that guides them.

The four elements theory went back to the Greek thinker Aristotle, who stated that the world was made up of air, fire, water, and earth and that these substances could be changed into each other under the right conditions. Lavoisier had already proven that water does not turn into earth.

Another theory was connected to how fire changes substances. The phlogiston theory, as it came to be called, began with Joachim Becher, a German scientist. Georg Ernst Stahl, a professor in Germany who lived from 1660-1734 greatly

Georg Ernst Stahl

Georg Ernestus Stahl, Onoldo Francus,
Med. Doct. h.t. Prof. Publ. Ord. Hall.

expanded the theory. Stahl believed that air was inert and did not participate in chemical reactions. As he studied combustion, he proposed that a substance he called phlogiston was involved in the process.

Stahl described the phlogiston theory like this: when a substance burns or is heated, it loses phlogiston. For example, when charcoal burns, there is very little ash left, so charcoal has a large amount of phlogiston that is released as it burns. When the metal lead is heated, a powdery material forms on the surface. The powder is known as calx. Stahl decided that metal must be a combination of calx and phlogiston. When the calx is heated with charcoal, it becomes the metal lead again. So the calx has absorbed phlogiston

The phlogiston theory sought to explain what happens when something, such as charcoal, burns.

from the charcoal and become lead. In short: Metal = calx + phlogiston.

One of the problems with the phlogiston theory was that metals gain weight when a calx is formed. Some scientists ignored this fact for years, claiming that most metals did not gain weight during calcination. Others said that phlogiston must have a negative weight, so as the metal lost phlogiston during calcination, it weighed less.

The mysterious phlogiston theory bothered Lavoisier. Still for the time, he accepted it.

Robert Boyle had shown that high heat could destroy diamonds, the hardest known substance. But scientists were not sure exactly what happened to the diamonds. Did they evaporate into the air or did they burn?

Lavoisier, Pierre Macquer, and Louis Cadet heated some diamonds in an open vessel for three hours. They found that the diamonds didn't shine anymore and they lost weight. A jeweler was certain that diamonds would not evaporate without air, so he gave them three more diamonds for the experiments. The diamonds were weighed carefully and sealed in a container.

For heat, the group resurrected a large burning lens that belonged to the Academy. The lens had been built in the early 1700s but had not been used for years. The contraption had a large lens made of hemispheres of glass, sixteen feet in diameter. Another movable lens could be used to more closely concentrate the sun's rays. The whole thing was mounted on wheels and had a series of levers so it could be moved to catch the sun. The burning lens, like a magnifying glass, focused the sun's rays on the container. Operators wore dark glasses to protect their eyes.

Engraving of the large burning lens Lavoisier used to burn diamonds.
(Courtesy of The Print Collector/Alamy)

Lavoisier and the other scientists set the burning lens up in the Jardin de l'Infante, the garden between the Louvre and the Seine River. Crowds gathered to watch the scientists trying to burn diamonds.

The scientists found that the diamonds did not lose weight if they were heated in the sealed container. Even the high heat of the burning lens made no difference. The diamonds only lost weight when they were heated in air. These results made Lavoisier more curious about the process of burning.

Where did the diamond go when it burned in air? How could the strongest known substance on earth be burned? And why did it not disappear when it was heated without air?

Lavoisier suspected that when a diamond was heated, a gas might be released. When he tried to burn diamonds in his lab under a glass bell jar and collect the gas, his glass vessels

Lavoisier and his colleagues performed their experiments in the gardens between the Louvre and the Seine.

broke. He abandoned that project. But the process of burning intrigued him and he performed experiments on other materials. Today it is known that diamonds are a form of carbon and when subjected to immense heat, the carbon burns and becomes carbon dioxide.

On August 19, 1772, Lavoisier presented a paper to the Academy of Science called "Memoir on Elementary Fire." He had written up the same lecture with the long title, "Reflections on Experiments Which One Might Try With the Aid of the Burning Glass." Obviously, he had many more ideas for the burning lens.

He began the paper by mentioning the theory of Stahl on phlogiston and the experiments by Frenchman M. Geoffroy, also done with the burning lens. Geoffroy had decided that all metals or metallic materials contain an earthy substance and an oily material. He stated that this oily matter, found in plants, animals, charcoal and metals, can be taken "out of them, and put . . . back into them at will."

Lavoisier stated that these two theories were the same. Stahl and Geoffroy had just used different names: phlogiston and oily material. Both were vague and mysterious. Lavoisier wrote "It is easy to see that this System does not differ from that of Stahl except in that M. Geoffroy calls oily Matter of inflammable Substance that which M. Stahl names phlogiston; well, it must be confessed that even today we don't yet know the nature of what we call phlogiston well enough to be able to declare anything very precise About Its nature." For a scientist who believed in facts proven by experimentation, phlogiston was too vague and added nothing to the understanding of chemistry.

His paper also contains a section "On Fixed Air, or rather, on the air contained in bodies." He wrote, "It seems constant that air enters into the composition of most minerals, even metals and in very great abundance. No chemist, however, has yet made air enter into the definition either of metals or of any mineral body." Although Lavoisier didn't know it, scientists in Scotland and England had already advanced beyond him in the study of what are today called gases.

In 1756, sixteen years before Lavoisier's experiments with the burning lens, a Scottish chemist, Joseph Black, had prepared a gas which he called "fixed air." The gas formed when chalk dissolved in acid. Black thought the air or gas was fixed in the chalk until it was released by the acid. He determined that fixed air was different from common air because it did not allow combustion, it killed small animals, and it was formed when chalk was dissolved in acid. Today fixed air is known as carbon dioxide.

Black also learned that fixed air was produced in the burning of charcoal, in respiration and in fermentation. In one experiment, he set up equipment in a church ceiling. He tested the air breathed out by a congregation of 1,500 people during a ten-hour church service. His experiment showed that the gas exhaled by his unknowing subjects was fixed air—carbon dioxide.

Another Englishman, Henry Cavendish, had produced a gas in 1766. He found that the gas burned, but was not the same as common air. He called it "inflammable air." Lavoisier would study this gas in the future.

Lavoisier didn't want English chemists to announce a theory before he could. He wanted priority of the ideas that were developing in his mind. So he scribbled out a note explaining some of his thoughts and had the secretary of the Academy sign it to establish the date. His note shows that he was struggling to devise a theory that would unite the things he knew about fire and air and the ideas he had not yet had time to prove by experimentation. He still accepted the phlogiston theory, stating that "fire enters into the composition of all Bodies." He also had a glimmer of understanding of a connection between the action of air and of fire. He wrote, "What we will say concerning air is equally true of phlogiston or the matter of fire."

But he realized that his work raised more questions than it answered. He wondered how air that could be so expansive, could attach to solids, and occupy a much smaller space. How can it be the same substance? "The solution to this problem tends toward a singular theory which I am going to try to make understood that is the air we breathe is not at all a simple being. It is a particular fluid combined with the matter of fire—" At this point, his short note ends.

To learn more about the mysterious phlogiston, he designed experiments with other substances that burned. He bought some phosphorus, a chemical that burns so easily in air that it must be stored in water to keep it from bursting into flame. On September 10, 1772, Lavoisier wrote in his notebook that he wanted to see if phosphorus absorbed air when it burned. For these experiments, he relied on equipment invented by Englishman Stephen Hales.

A reproduction of Lavoisier's lab

Hales had expanded Robert Boyle's apparatus for collecting gases from an experiment. The equipment had a chamber where a substance could be heated. A tube leading from the chamber would carry gases released into a vessel turned upside down in a bowl of water. The gas would collect at the top as the water level in the vessel dropped. Hales had reported that a large amount of air was absorbed when phosphorus was burned. Lavoisier planned to repeat his experiments.

Lavoisier found that phosphorus did absorb a large quantity of air when it burned. He also found that an acid was produced in the reaction. Another scientist and friend of Lavoisier's, P. F. Mitouard, had discovered that the acid that was produced weighed more than the original phosphorus. Lavoisier wanted to find if the extra weight came from water in the air or from the air itself.

One challenge was determining the weight of the acid produced. In the experiment, the acid dissolved in water. So how do you weigh something in water? Lavoisier remembered his expedition with Guettard to study geology and water sources around France. One of their instruments was a hydrometer, used to measure the density of water. A sample of water with a dissolved mineral would weigh more than a sample of pure water.

Lavoisier used this method in determining the weight of acid produced. He burned bits of phosphorus and collected the acid produced in a beaker. He meticulously weighed the beaker and marked the level of acid on the side. Then he poured out the acid and filled the beaker with water to the same mark and weighed it. The difference in weights was the amount of acid from the experiment.

By carefully weighing the phosphorus and the acid, Lavoisier found that the phosphorus gained more than a third of its weight when it burned. He decided to see if the same thing happened when other substances were burned.

He next burned sulfur and found that it also produced an acid when it was burned. Again carefully weighing all the parts of the experiments, he found that sulfur gained weight just as the phosphorus had. Other scientists had performed these experiments and gotten similar results. But Lavoisier realized what was happening. In a sealed report to the Academy on November 1, 1772, he wrote "this increase of weight arises from a prodigious quantity of air that is fixed [attached] during the combustion." In other words, the extra weight comes from air attaching to the phosphorus and the sulfur. His sealed note was kept by the Academy until he requested it to be opened and read in May 1773.

Lavoisier took an insightful leap from the results with phosphorus and sulfur to the process of heating metals. When a metal, such as lead or tin, is heated, a powder forms on the surface. The powder is called calx and the process is called calcination. Iron forms a calx without heat. It is the red powder commonly known as rust.

When a metal calx was heated with charcoal, it released a gas, the charcoal disappeared, and the metal was reformed. Lavoisier determined to know more about this process and if it applied to all metals. He also wondered what gas was released.

In his November 1 note, Lavoisier wrote that he heated lead calx and a "large quantity of air was liberated and that this air formed a volume a thousand times greater than the quantity of litharge [lead calx] employed. This discovery appearing to me one of the most interesting of those that have been made since the time of Stahl, I felt that I ought to secure my right to it, by depositing this note in the hands of the Secretary of the Academy, to remain sealed until the time when I shall make my experiments known." Before he announced the discoveries he wrote about in the November note, Lavoisier planned to perform many more experiments. But he ran into problems with his equipment. The bell jars covering a metallic sample often broke when the burning lens was focused on them. In February 1773, a glass jar broke when he used a furnace to heat a sample of lead calx in it. Lavoisier designed his own glassware, but workmen were not able to make them in time for his experiments. So he put together items he could find around the house; ordinary glass jars, a wash basin, and a crystal pedestal normally used to serve fruit.

The glass problems delayed his experiments, but he still wanted to present some results to the Academy in the spring of 1773. He wrote in a notebook that other scientists had made great advances in the study of gases, acknowledging that Black and others had "observed differences so great between the air liberated from substances and that which we breathe, that they deemed it to be another substance, to which they have given the name of fixed air."

He thought that more work was needed to determine exactly what the gases were. The experiments had "come far short of the number necessary for a complete body of doctrine." Lavoisier planned on using the study of gases as a foundation for a new system of chemistry. He knew that the work had to be based on those who had come before:

> The importance of the end in view prompted me to undertake all this work, which seemed to me destined to bring about a revolution in physics and chemistry. I have felt bound to look upon all that has been done before me merely as suggestive:

I have proposed to repeat it all with new safeguards, in order to link our knowledge of the air that goes into combination or is liberated from substances, with other acquired knowledge, and to form a theory. The results of the other authors whom I named . . . appeared to me like separate pieces of a great chain; these authors have joined only some links of the chain. But an immense series of experiments remains to be made in order to lead to a continuous whole.

In April, Lavoisier presented his findings to the Academy. He still had not been able to finish the experiments he planned, but he had thought through the process enough to see where he was headed. His conclusion:

It obviously results from these experiments, 1st that a metallic calx is nothing other than the metal itself combined with fixed air, 2nd that the metallic reduction consists only of the disengagement of the air from metallic calces, 3rd that the metals owe the weight gain to the fixed air contained in the atmosphere."

He realized that his conclusion contradicted the theory of phlogiston. He wrote, "above all decisive experiments have assured me that it is possible to reduce almost all metals without the addition of phlogiston. . . . I have even come to the point of doubting if what Stahl calls phlogiston exists, at least in the sense that he gives to that word, and it seems to me that in every case one could substitute the name of matter of fire, of light, and of heat."

Now that he had begun the attack on the phlogiston theory, he asked for the note he had sealed on November 1, 1772, to be opened and read for the Academy. By this action, he displayed his right to his discovery about metallic calces and air.

Lavoisier demonstrates his work to other scientists in his lab, while Marie Anne looks on. *(Courtesy of The Print Collector/Alamy)*

Other scientists, including American Benjamin Franklin, doubted Lavoisier's ideas at first. Franklin, in France to campaign for support from the French government for the American Revolution, had become a good friend of the Lavoisiers. He often ate at their house and gathered with other scientists from around the world. Franklin wrote another scientist friend, Jean-Baptiste LeRoy of the French Academy of Sciences, "I should like to hear how M. Lavoisier's doctrine supports itself, as I suppose it will be controverted."

As Lavoisier continued to perform experiments in an attempt to strengthen his theory, he ran into more problems. He believed that common air was a mixture of gases and that only part of the air reacted with metal to form the calces. Lavoisier realized that the calcination of metals stopped when the part of air was used up. However, his results were often confusing. Today it is known that sometimes his experiments produced carbon dioxide and sometimes the product was oxygen.

Lavoisier collected his experiments and results into his first book, *Physical and Chemical Essays.* He put it together in late 1773 and it was published the next year. However, 1774 would bring confirmation of another discovery from England that piqued Lavoisier's interest.

four

The Oxygen Theory

The Scottish chemist Joseph Black had discovered fixed air in 1754. Another discovery was that fixed air (carbon dioxide) would form a precipitate in limewater. If the gas was passed through limewater, the solution turned milky white. This test could be used to identify fixed air produced in reactions.

Joseph Priestley, an English chemist and minister, was also conducting experiments on air. Priestley had met Benjamin Franklin in London in 1766 and the meeting inspired an interest in science. As librarian and secretary to William Petty, second Earl of Shelburne, he had time from his duties to experiment.

Living near a brewhouse in Leeds, England, Priestley's curiosity was aroused by the fermentation process. He devised a process to collect the gas released in the brewing process—

Black's fixed air—and dissolve it in water. Thus, soda water was invented.

Priestley improved Hale's pneumatic trough by using mercury instead of water, allowing for the collection of water-soluble gases. His most important discovery was of a gas that he named "dephlogisticated air."

He heated red calx of mercury (mercuric oxide) with a lens and captured the gas released. He found that it was insoluble in water and supported respiration and combustion. Priestley, along with other chemists, thought that phlogiston was released during respiration and combustion and absorbed by the air. Since both respiration and combustion ceased if fresh air was not available, they thought that the air had absorbed as much phlogiston as possible, much like

Joseph Priestley

a sponge that soaks up water. When Priestley found that a candle and a stick of wood burned vigorously in the new air (gas), he thought the gas must be free from phlogiston. The new gas absorbed an incredible amount of phlogiston, therefore, he called it dephlogisticated air.

Lord Shelburne and Priestley set off on a European trip, so experiments ceased for a time. While in Paris in October 1774, Priestley visited with Lavoisier and other scientists at Lavoisier's house. He told them about the exciting new air and its properties.

After returning to England, Priestley continued his experiments. On March 8, 1775, Priestley put a mouse into the new air and observed that it seemed to do well. He concluded that this new air must be even better than common air and predicted that it might be useful for people with lung problems. Priestley decided to breathe the new air himself. He wrote, "The feeling of it to my lungs was not sensibly different from that of common air; but I fancied that my breast felt peculiarly light and easy for some time afterwards. Who can tell but that, in time, this pure air may become a fashionable article in luxury? Hither to only two mice and myself have had the privilege of breathing it."

Priestley is considered the discoverer of the gas known today as oxygen. Swedish scientist Karl Scheele also prepared what he called "fire air" but he did not publish his findings until after Priestley had announced his discovery. However, it was up to Frenchman Antoine Lavoisier to describe what really happened to oxygen in reactions and eventually, to name it.

A disagreement broke out between two French chemists, and Lavoisier was part of a committee selected to resolve the

Karl Scheele experimented with a gas he called "fire air."

controversy. Antoine Baumé stated that red mercury calx could not be reduced to mercury without the phlogiston from charcoal, while Louis Cadet said that it could be done. The committee performed experiments and proved that Cadet was right; mercury calx could be reduced to mercury by heating without charcoal. None of the scientists seemed to pay attention to the gas that was released.

However, after Priestley's visit, Lavoisier returned to the experiments with red mercury calx. In March 1775, he deposited a sealed note with the secretary of the Academy stating his findings and conclusions.

Antoine Baumé

He had heated red mercury calx with charcoal and tested the gas released. He found it dissolved in water, precipitated limewater, and prevented combustion or respiration. In other words, it was fixed air (carbon dioxide).

When he heated the calx without charcoal, he obtained a different gas. This one did not dissolve in water, or precipitate in limewater, and it did support combustion and respiration. Lavoisier called this gas, "eminently breathable air."

He did not grasp yet that it was not just air, but a component of air. But he was on his way to that conclusion.

Priestley criticized Lavoisier's results, stating that Lavoisier had not carried out enough experiments to prove what he postulated and that Priestley himself had performed many of the same experiments. Priestley hit on a problem in Lavoisier's ideas: when calces are formed, they absorb only part of the common air, not the whole. Lavoisier extended his experiments to understand this discrepancy.

Lavoisier determined that when mercury calx was formed, part of the air did not participate in the reaction. He called this component of the air mofette; today it is called nitrogen. Lavoisier could separate air into its component parts; eminently breathable air (oxygen) and moffette (nitrogen) and could combine them back into common air. He also determined that when mercury calx was reduced with charcoal, a different gas was formed, fixed air as discovered by Black. Lavoisier saw that adding charcoal into the reaction changed the gas released. The fixed air was a product of the charcoal and the eminently breathable air.

When chemists study a compound, they want to carry out the reaction in both directions; both synthesizing and decomposing a product. Lavoisier carried this out with air. "Here is the most complete sort of proof at which one may arrive in chemistry: the decomposition of air and its recomposition."

Then Lavoisier took a tangent in his experiments that led to the name for eminently breathable air. He began to investigate the production of acids. When phosphorus is burned, it combines with air and produces "acid of phosphorus." Burning sulfur also produces an acid. In Lavoisier's time it was called vitriolic acid; it is today called sulfuric acid. He knew that, in burning, the phosphorus and sulfur combined with eminently breathable air, so he concluded that all acids contained the same material. He called this the acidifying principle. On September 5, 1777, he reported to the Academy that he had named this substance. "I shall henceforward designate dephlogisticated air or eminently respirable air in the state of combination and fixity, by the name of the acidifying principle, or if one likes better the same meaning in a Greek word, by that of *le principe oxygine* [the principle oxygen]."

So Priestley and Scheele had discovered the gas that Lavoisier now named oxygen. The name oxygen comes from Greek words meaning acid maker. Lavoisier thought that all acids contained oxygen. He was wrong in that, as later experiments would show.

At the same time, Lavoisier was growing more dissatisfied with the phlogiston theory. He saw that his explanation of oxygen combining with metals to form calx and of its action in respiration contradicted the theory and a new one was needed to take its place. He wrote, "Besides, since I am at the point of attacking the entire doctrine of Stahl concerning phlogiston, and of undertaking to prove that it is erroneous in every respect, if my opinions are well founded M. Priestley's phlogisticated air will find itself entangled in the ruins of the edifice."

As he continued experiments and began to develop the oxygen theory as explanation for combustion, Lavoisier used phrases like "matter of fire" less. He began to use the word "caloric." (Changed to calorie over time: it is a measure of heat.)

Lavoisier was aware of Joseph Black's theory of latent heat. Black had stated that water absorbed an amazing amount of heat during melting. If ice melted all at once, floods would occur every spring. Instead ice and snow melt gradually. Black's experiments showed that the temperature of melting water was stable until the ice had all melted. In his experiments, Black froze an amount of water in one tube and cooled the same amount to almost freezing in another tube. Then he placed them in a warm room. The cooled tube started at 33° F and in half an hour the temperature rose to 40°. The frozen tube, starting at 32° took ten and a half hours to reach 40°.

An illustration from the 1700s depicts Lavoisier studying the gas he will eventually name "oxygen." *(Courtesy of North Wind Picture Archives/Alamy)*

Since that was twenty-one times longer (630 minutes related to thirty) than the cooled water had taken, Black assumed that twenty-one times more heat had entered the ice.

Later, Black showed that an even larger amount of heat was required to turn liquid water into steam. His work on what is called latent heat cleared the way for work by James Watt on steam engines and advanced scientific ideas about heat.

In 1782, Lavoisier and his friend, Simon de Laplace— known for his work in mathematics, but also an excellent designer of scientific instruments—used Black's ideas to develop a calorimeter. The tool had three nesting chambers. Two were filled with ice and the innermost one held the components of the experiments. Whatever heat was produced in the inner chamber would melt the ice of the second one, giving a measurement by the amount of water formed.

In earlier experiments, Lavoisier had discovered that oxygen in the air turned to carbon dioxide in the lungs. He hypothesized that an animal maintains body heat by release of matter of fire in the lungs as the oxygen changes to carbon dioxide. Now he planned experiments to prove his hypothesis.

Lavoisier and Laplace measured the amount of carbon dioxide produced by burning a weighed amount of charcoal and also the amount of heat produced. These two numbers led to an amount of heat produced in relation to the amount of carbon dioxide produced. Then they put a guinea pig in the chamber of the calorimeter for ten hours and measured the amount of ice melted. They found that the heat of the guinea pig's body melted close to the amount expected from the amount of carbon dioxide that it exhaled. So the conversion of oxygen to carbon dioxide in the guinea pig's lungs provided the body heat. Today it is known that what happens is more

Pierre Simon de Laplace

complicated than Lavoisier theorized, but the results obtained by him and Laplace were close to what happens.

These experiments led Lavoisier to conclude that respiration is a slow form of combustion. "Thus the air that we breathe serves for two purposes equally necessary for our preservation: it takes away from the blood the base of fixed air [carbon dioxide], the excess of which would be very harmful; and the heat that this combination sets free in the lungs makes up for the continual loss of heat that we experience with regard to the atmosphere and surrounding bodies."

In spite of the political upheaval of the French Revolution in 1790, Lavoisier continued conducting scientific experiments. He and Armand Séguin studied human respiration. Séguin, as human guinea pig, wore a mask that collected his respiration. Lavoisier also measured Séguin's breathing rate and pulse.

Marie Lavoisier made drawings of the experiments. She drew Lavoisier attending the pneumatic trough which captured the gas produced by Séguin's respiration and she drew herself as she worked on the drawing or notes.

Lavoisier tested Séguin's respiration as he sat still and as he performed work by pushing a lever, which raised a weight, with his foot. He also tested Séguin after he had eaten. The experiments showed that Séguin used almost three times more oxygen when he worked than sitting still. His breathing rate and pulse also accelerated during work. Lavoisier also found that more oxygen was used while digesting food, or on cold days. These results supported Lavoisier's theory that oxygen consumed in respiration helped to heat the body.

Lavoisier had said in earlier papers that respiration was a combustion process. In a report to the Academy of Sciences, Lavoisier said the fuel for the combustion came from food. He compared food to the oil of a lamp. "If the animal did not receive . . . from food what it loses by respiration, the lamps would soon run short of oil. The animal [would] perish, as a lamp goes out when the fuel is exhausted."

Lavoisier conducted other experiments that convinced him that living tissue, whether plant or animal, was predominantly composed of the elements carbon, oxygen, hydrogen, and nitrogen. Today, the field of organic chemistry is defined as the study of carbon and its compounds.

However, there were still holes in Lavoisier's oxygen theory. To fill the holes, he needed information from another English chemist.

Henry Cavendish had discovered inflammable air, what is known today as hydrogen. The gas formed when vitriolic acid [sulfuric acid] was poured on iron. Cavendish determined that burning inflammable air in common air would decrease the volume of the air by one-fifth. He mentioned that the airs "condensed into the dew which lines the glass." However, Cavendish did not state that the hydrogen and oxygen had combined to form the dew. Priestley had also burned hydrogen in oxygen and noticed water, but he failed to realize where the water had come from.

Cavendish had reported his experiments in 1781, but French scientists did not know about them until a visit in 1783 by Charles Blagden, Cavendish's assistant. Blagden met Lavoisier and in swapping experiments and tales, he told Lavoisier about the dew. Lavoisier needed to investigate this phenomenon. While Blagden was still at his laboratory, Lavoisier and Laplace performed the experiments. The next day, Lavoisier made a brief report to the Academy stating that burning inflammable air with oxygen produced "water in a very pure state."

Lavoisier's experimental results were not as good as Cavendish's. He reported that the weight of water obtained was almost the amount lost by the two gases. However, Lavoisier made the theoretical jump to the fact that the water came from the combustion of hydrogen. At a public meeting of the Academy, he read a report, "On the Nature of Water and on Experiments that appear to prove that this Substance is not properly speaking an Element, but can be decomposed and recombined."

To prove that water was not an element, Lavoisier needed to decompose it and to form it from the two gases, hydrogen and oxygen. He introduced water into a glass bowl suspended over mercury. The bowl also contained iron filings which gradually rusted; the iron combining with oxygen from the water. As this process occurred, Lavoisier collected the gas that was released. When it was tested, it was hydrogen. Therefore, he had made water by combining hydrogen and oxygen and he had decomposed water by splitting it into oxygen and hydrogen. He said, "water is not an element, it is on the contrary composed of two very distinct principles, the base of vital air [oxygen] and that of hydrogen gas; and . . . these two principles enter into an approximate relationship of 85 to 15 respectively."

Another of the four elements of the Aristotelian system was proven to be a compound. By now, Lavoisier had shown that earth, air, and water were not elements. Only fire remained of the four. He would soon take on the character of fire in attacking the phlogiston theory. In the meantime, Lavoisier was also working in various capacities for the government of France and for the Academy of Sciences.

five
Government Work

While Lavoisier seemed to be happiest working in his laboratory, he had to limit his time there. His usual routine was to work in the lab from six until nine in the morning and from seven to ten in the evening. One day a week, he spent the entire day in the lab and according to Madame Lavoisier, "It was for him a day of happiness; some friends who shared his views and some young men proud to be admitted to the honor of collaborating in his experiments assembled in the morning in the laboratory."

The rest of the hours of the week were devoted to Lavoisier's other responsibilities. In 1768, he had joined the Tax Farm and through the years, he campaigned to have taxes collected in a more just method. The organization was divided into committees. Lavoisier began by serving on the tobacco committee, charged with making sure tobacco sold to the public was not adulterated with ashes. By adding ashes, storeowners added

An engraving depicting Lavoisier experimenting with oxygen and hydrogen to produce water. *(Courtesy of Pictorial Press Ltd/Alamy)*

weight and volume to their product. Cheating in this way aroused Lavoisier's sense of fairness and he determined to find a way to distinguish ashes in tobacco. "Happily chemistry provides a reliable and unequivocal test that enables one to detect ashes in substances with which they have been mixed." The method was to drop acid on the tobacco. If ashes, which are alkaline, were present, the mixture of acid and alkaline would effervesce.

One problem faced by the Tax Farm and by extension, the government, was smuggling to avoid paying taxes. It was estimated that as much as one-fifth of the merchandise was brought into Paris by smugglers. Therefore, the Tax Farm and the government lost the tax revenue on those goods. Lavoisier proposed that a wall be built around Paris. People and supplies would only be able to enter the city through gates. In 1783, the government decided to implement this proposal. They chose an architect to build the structure. Unfortunately, he built elaborate buildings at the gates, costing a lot of money and earning the criticism of the people. Some said that the Tax Farm was trying to keep fresh air out of Paris by building the wall and others claimed that the purpose of the wall was to make the Tax Farmers rich. Lavoisier was blamed for this fiasco and unjustly portrayed as one who didn't care about the common people.

For his service and investment in the Tax Farm, Lavoisier received an income, a share of the company profits, and interest on the money he had invested. It was not a fortune but this pay and money from other jobs, as well as the wealth he inherited, allowed Lavoisier to continue his scientific research. Most citizens, however, thought of the Tax Farmers as financiers who grew wealthy on the taxes extorted from the people.

The Rotunda of the Barrier of La Villet is one of the only portions of the Tax Farm's walls that remains today.

King Louis XV died in 1774 and his son, Louis XVI, became king. He appointed A. J. M. Turgot as his controller-general. Turgot believed that the government of France needed economic reform. He asked the new king to not spend the money for an elaborate coronation, especially since the people were suffering from a high price for grain. The king celebrated in grand style regardless of Turgot's advice.

Lavoisier agreed with Turgot that economic and tax reform was necessary. Soon after Turgot came to power, he appointed Lavoisier to a new job: gunpowder commissioner.

The supply of gunpowder had been in control of a private company organized much like the Tax Farm. However, the company made little powder and the quality was very poor. French soldiers could not count on the powder's effectiveness and they sometimes lost battles because of poor gunpowder for their guns. The government even imported gunpowder from other countries to have a supply for the army. The imported powder was expensive and its purchase took more money out of the strained national treasury.

As a chemist, Lavoisier was uniquely qualified for the job of gunpowder commissioner. With the position came living quarters at the Paris Arsenal where the gunpowder was kept. The Lavoisiers moved into spacious living quarters and Antoine set up an elaborate laboratory there. He began investigating the manufacture of gunpowder.

The components of gunpowder are sulfur, charcoal, and niter (potassium nitrate). The niter comes from a compound called saltpeter. Saltpeter forms naturally in damp places and near animal pens. Under the control of the old company, saltpeter collectors had the right to search anyone's property for saltpeter and then dig it up with little compensation to the landowner. They also could stay in a town for free and buy wood needed to make saltpeter at a very low cost. Often, they took bribes. These saltpeter inspectors were disliked as much as the tax collectors.

Lavoisier read about people making their own saltpeter in niter beds. The process involves digging ditches and filling them with manure, plant material, and chalky dirt. Water was added and after a few months, wood ashes were mixed in and the substance was leached with hot water. After the hot water evaporated, saltpeter remained.

Lavoisier proposed that the French also make their own saltpeter. When his program was implemented, the amount and quality of saltpeter and gunpowder increased significantly. In 1755, the gunpowder supply was less than half of what France needed. By 1776 and 1777, France was supplying gunpowder to the American colonies fighting the British in the Revolutionary War. Lavoisier wrote, "One can truly say that North America owes its independence to French gunpowder." The quality improved so much by 1788 that a cannonball could go twice as far with the new powder.

During this research on gunpowder, Lavoisier was almost killed. The chemist, Claude-Louis Berthollet, discovered that mixing a salt called muriate of potash with carbon produced more powerful gunpowder than that made with saltpeter. Lavoisier and his wife and others traveled to the country where the powder was going to be ground up and tested. A barrier had been built to protect the spectators, but another gunpowder commissioner, Le Tort, was in front of the barrier, trying to speed up the process. At a time when Lavoisier, Madame Lavoisier, and the other spectators were a distance away, the powder mill exploded, killing Le Tort and a woman.

In 1775, Lavoisier's work on gunpowder was temporarily halted by the death of his father. Grieving for his father, whom he always had been close to, Lavoisier wrote, "It is less the loss of a father that I mourn, than the loss of my best friend."

In 1776, Turgot was forced out of office. For a time, Lavoisier was afraid he would lose his job as gunpowder commissioner, but he was able to keep the job and his laboratory at the Arsenal.

Besides his duties with the Tax Farm and the Gunpowder Commission, Lavoisier became interested in agriculture. In 1778, he bought a farm named Fréchines and this land became the first experimental farm in Europe. Lavoisier brought his powers of scientific analysis and measurement to the farm.

Lavoisier believed that animals and crops are both equally important to farms. Manure from animals can be used to fertilize crops and crops used to feed animals. Hay was an important crop to provide wintertime food for animals. Lavoisier set out to prove to farmers that his methods would work. He joined the Society of Agriculture and encouraged others to carry out reforms in farming.

Seed, fertilizer, and the yield of crops were carefully weighed. Lavoisier also instigated crop rotation and a new crop, potatoes. The results of Lavoisier's farm experiment would take years to obtain. But by 1793, the farm was yielding twice as much wheat as when he bought it. George Washington wrote for Lavoisier's advice and put it into action on his farm in America, Mount Vernon.

When a severe famine hit the town of Blois, Lavoisier advanced no-interest loans to the town to buy wheat for the starving. The town named him an honorary citizen saying, "Lavoisier . . . is a great and good man who saved our farming district."

Lavoisier believed in balance and careful measurements in science. He carried that belief into his farming ventures and into economics. Lavoisier agreed with a group of economists known as physiocrats, led by Turgot. These men believed that agriculture was the most important industry and must develop economically as much as possible. They advocated free trade between the provinces of France and a uniform system of taxation. The current system of taxing trade between provinces and of the wealthy not paying taxes, while the poor struggled to pay taxes and feed a family, were dragging the country down.

Lavoisier helped the people of the town of Blois buy wheat during a famine. *(Library of Congress)*

In a report he wrote, Lavoisier stressed the danger of controls on trade. Some farmers were killing lambs and calves because there wasn't enough to feed them. The province wanted to prohibit the practice. Lavoisier wrote, "Increase in price decreases consumption, accelerates the return of plenty, and acts as a premium to attract supplies even from foreign sources, and such a premium has the advantage that it costs the State nothing." In other words, balance and equilibrium in economics were as critical as in other parts of his life.

As early as 1784, he had begun making a calculation of the gross national product of France. He advocated precise accounting of money in every area of public life. He saw money as "a fluid whose movements necessarily result in equilibrium." If money stopped circulating, because people were hoarding it, or if there wasn't enough due to the excesses of the court, the country and its people suffered.

In 1788, Lavoisier became a director of the Discount Bank. This was a private institution that existed to loan money to the French government. As the court spent more and more money and as unsettled political times approached, Lavoisier and the other directors struggled to keep the bank and the Royal Treasury solvent.

Lavoisier also had assignments and duties with the Academy of Science. For almost every committee he was on, Lavoisier was the secretary. He wrote the reports that each committee published.

He studied prisons and hospitals. Both places were dirty and full of disease. Lavoisier reported that the prisons were extremely crowded, full of sickness and filth. Prisoners of all types, from those convicted to crimes to the innocent accused of crimes, were housed together.

Lavoisier's report recommended buildings with good ventilation and fresh water be constructed. He also thought that prisoners should be separated by crime and whether or not they had been convicted. Clothing and some basic furniture should be provided as well as space to sleep. The recommendations in the report were never instigated.

Hospitals were not much better than the prisons. When Lavoisier wrote a report to the Academy in 1787, he found that patients were crowded into beds, regardless of their condition. A smallpox patient might be in the same bed as a fever patient. Contagious diseases spread rapidly through the entire hospital. The noise level was appalling, so rest was impossible.

An Austrian doctor, Anton Mesmer, claimed to have developed a method of curing people involving what he called

Anton Mesmer

"animal magnetism." The process, mesmerism, was named for its inventor. The Academy of Science formed a committee to investigate Mesmer's methods. The scientists were concerned that the method might not only be a hoax, but even detrimental to people. The committee chose to investigate included Lavoisier, Benjamin Franklin, and Dr. Joseph-Ignace

Dr. Joseph-Ignace Guillotin, the inventor of the guillotine, helped Lavoisier investigate Mesmer.

Guillotin, inventor of the eponymous execution device.

Magnetism was known to the people of the time. It is the force that causes compass needles to point north. Mesmer stated that animal magnetism could be found in people and passed from one person to another, curing diseases. When people thought they had been magnetized, they often had fits or fell into trances.

Lavoisier knew that it would be difficult to determine if Mesmer's method was curing people. He wrote "Nature, left to her own devices, cures a great many diseases; and when remedies have been applied, it is extremely difficult to decide how much is due to Nature and how much to the remedy."

When the committee was investigating the method, Mesmer was not in Paris, but his assistant Charles Deslon was treating people. The mesmerism treatments were similar to a séance. With music playing in the background, people sat around tubs full of water bottles that were supposed to be magnetized. Iron rods stuck out of the tubs. People held hands and sometimes touched the iron rods to parts of the body that needed healing. Deslon would walk through the room, waving a wand to magnetize people.

Lavoisier and his colleagues brought instruments to measure electricity and magnetism. The instruments were not able to detect anything in the water, the rods, or Deslon's wand. They blindfolded people and told them that Deslon was magnetizing them; many of these people had fits, even when Deslon was not there. Other times, they had people sit behind a screen while Deslon tried to magnetize them. The people did not know they were being "treated" and they showed no reaction. Benjamin Franklin submitted to the treatments in the name of science but had no change from the process.

The committee published a report stating that animal magnetism did not exist. People reacted because they believed in it and if they were healed, it was the effect of their own minds. "We discovered we could influence them ourselves so that their answers were the same, whether they had been magnetized or not," Franklin wrote. He also said, "Some think it will put an end to Mesmerism, but there is a wonderful deal of credulity in the world, and deceptions as absurd have supported themselves for ages."

The investigation and report on mesmerism built resentment among its practitioners against Lavoisier and the other discrediting scientists. Two of the believers were Jean-Paul Marat and Jacques-Pierre Brissot. The hatred felt by these men would have a huge impact on Lavoisier in a few years.

Another project investigated by Lavoisier was the "aerostatic machine" or balloon. The brothers Montgolfier had launched a balloon filled with hot air. More experiments led to higher altitudes and longer flights. The Academy of Science conducted a successful experiment and then planned a demonstration before the king and queen at Versailles on September 19, 1783. The balloon, carrying a sheep, a rooster, and a duck, ascended to more than 1,400 feet and traveled nearly two miles before it landed as the air cooled inside the balloon.

A model of the balloon made by the Montgolfier brothers.

On October 15, 1783, Pilatre de Rozier made the first human ascent in a balloon.

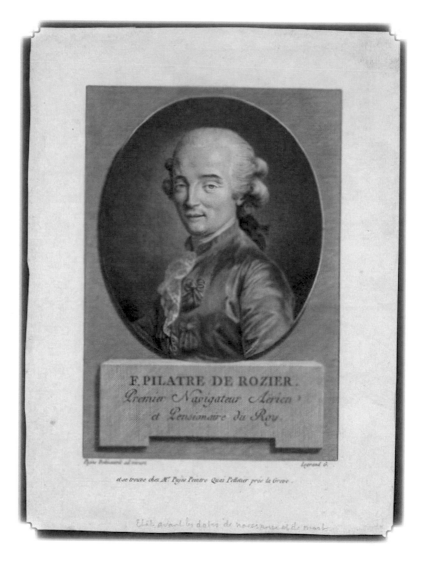

Pilatre de Rozier *(Library of Congress)*

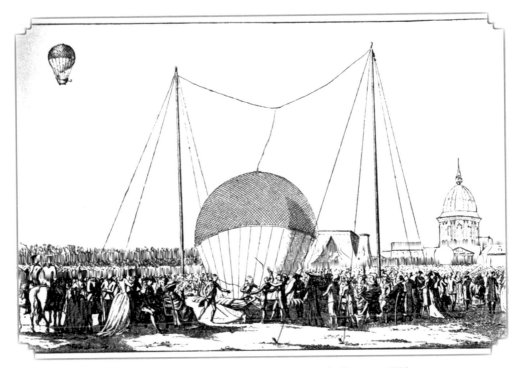

Pilatre de Rozier made the first ascent in a balloon in 1783.

Scientists began looking for alternatives for lifting the balloons. Hot air balloons would have to carry considerable fuel for heavy loads and long flights. Jacques Charles considered inflammable air—hydrogen. It was found to allow balloon flight without the extra load of fuel. However, a way to make a large amount of hydrogen was needed. Lavoisier began investigating ways to produce hydrogen, and again his mind returned to the problems with the theory of phlogiston.

six

Attack on Phlogiston

By the late 1770s, Lavoisier had definitively shown that earth, water, and air were not elements, destroying finally the Aristotelian theory. But another theory still held chemistry in its grasp: the phlogiston theory.

When Joachim Becher and Georg Stahl had proposed the idea of phlogiston, it fit the situation that they saw. In studies of combustion, they saw that some substances burned very easily and others not at all. So the theory said that when a material is burned or heated, it loses phlogiston. If the reduction process is followed, that is, metal calx heated with charcoal, the phlogiston is restored and the metal reappears. This process was valuable to the mining industry. Smelting requires heating the ore to obtain the purified metal.

According to the theory, a candle under a bell jar ceases to burn when the air is full of phlogiston. When Priestley isolated oxygen, he found that this gas allowed candles to burn brightly and for a long time, so he called it dephlogisticated air. Cavendish thought that the hydrogen gas that he discovered was pure phlogiston.

One big problem with phlogiston concerned the weight gain of a metal as it is heated and forms a calx. If it is losing phlogiston, how is it gaining weight?

In his early days, Lavoisier had accepted the phlogiston theory. However, he also wrote often about a substance he called "matter of fire." In 1766, he thought that air was water turned into vapor by matter of fire. For a while, he tended to confuse vaporization and evaporation with air. His political colleague, Turgot, invented the term "vaporize" and suggested that change of state from liquid to vapor occurred when matter of fire was added. Turgot also stated that all matter could exist in three states, depending on their content of matter of fire. Lavoisier agreed with him at this time. In a note, Lavoisier wrote that possibly air was an expanded fluid, a liquid that had been added to matter of fire.

Another question concerned effervescence. Lavoisier had noticed that effervescence could have a cooling effect. Did this mean that phlogiston or matter of fire was absorbed? Why did the temperature of melting water not increase until the water had melted?

Lavoisier had been attacking the phlogiston theory for some time, making small raids against it. In 1772, he said: "It is evident that Stahl's theory on the calcination and reduction of metals is seriously flawed and needs to be modified. It treats calcination solely as a loss of phlogiston, although it has been demonstrated that a loss of phlogiston and an absorption of air occur at the same time."

Lavoisier's oxygen theory had dealt a serious blow to the phlogiston theory by proving that air combined with metals to form calx. The metal was not the calx combined with phlogiston. As Lavoisier continued experiments, more and more of the phlogiston theory was proven false.

For a while Lavoisier thought that maybe fire could be fixed in materials just as air could. By 1777, he believed that matter of fire was not in the material that burned, but in the fluid surrounding the material. He wrote, "matter of fire is a very subtle and very elastic fluid that surrounds all parts of the planet we live on, which penetrates with greater or lesser ease all the bodies of which it is composed, and which tends, when free, to distribute itself uniformly in everything." The only difference between Stahl's phlogiston and Lavoisier's matter of fire seemed to be location. But by 1783, Lavoisier was ready for a full scale attack on the phlogiston theory.

In a memoir, "Reflections on Phlogiston," given before the Academy of Science in 1785, Lavoisier reviewed his past objections about phlogiston. He reminded the Academy of his experiments that led to the oxygen theory—that the processes of calcination, combustion, and respiration are all results of addition of oxygen.

He covered the history of Stahl's work and pointed out that he had made valuable discoveries. Lavoisier thought that Stahl's discovery of an inflammable principle in metals was exceptional. The realization that the property of combustibility could be transferred from one substance to another was a brilliant discovery. This statement came from the experiment of adding vitriolic acid [sulfuric acid], an incombustible material to charcoal, a very combustible

Lavoisier spent much time in his lab, studying fire and the phlogiston theory. *(Courtesy of Mary Evans Picture Library/Alamy)*

material. In this process, the incombustible acid converted to sulfur, a combustible material, and the charcoal lost its combustibility. Therefore, the combustibility was transferred from the charcoal to the vitriolic acid. From these experiments, Stahl concluded that a material known as phlogiston brought about these results.

Phlogiston was not a bad concept as far as facts that scientists of the time knew. But newly discovered facts disputed the presence of phlogiston. Lavoisier knew that those who accepted the existence of phlogiston had to face contradictions in their theory. He wrote,

> All these reflections confirm what I have advanced, what I set out to prove, and what I am going to repeat again. Chemists have made phlogiston a vague principle, which is not strictly defined and which consequently fits all the explanations demanded of it. Sometimes it has weight, sometimes it has not; sometimes it is free fire, sometimes it is fire combined with earth; sometimes it passes through the pores of vessels, sometimes these are impenetrable to it. It explains at once causticity and non-causticity, transparency and opacity, color and absence of color. It is a veritable Proteus that changes its form every instant!

For Lavoisier, who based his conclusions on the facts shown by experiment, the chameleon phlogiston had to go. "It is time to lead chemistry back to a stricter way of thinking, to strip the facts, with which this science is daily enriched, of the additions of rationality and prejudice, to distinguish what is fact and observation from what is system and hypothesis, and in short, to mark out, as it were, the limit that chemical knowledge has reached, so that those who come after us may set out from that point and confidently go forward to the advancement of the science."

Lavoisier listed the facts known about combustion and given clarity by his oxygen theory:

> 1. Combustion, including the formation of heat and light, is only possible in the presence of oxygen and when oxygen is gone, combustion stops.

> 2. All combustion occurs by an absorption of oxygen.

> 3. Every combustion is accompanied by a gain in weight of the body that burns and this weight is equal to the amount of oxygen consumed.

> 4. In every combustion, heat and light are produced.

His oxygen theory explained many of the same experimental results as phlogiston with an unambiguous precision that appealed to Lavoisier.

Lavoisier also included a section on his idea of heat. He hypothesized that "the molecules of bodies do not touch, that there is a space between them that heat increases and cold diminishes." Today it's known that heat is not the material substance that Lavoisier proposed, but more heat does increase the distance between molecules—added heat turns liquids into gases. However, the mechanism is that an increase in the movement or energy of the molecules causes the liquid to turn into gas.

In his conclusion, Lavoisier said,

> My only object in this memoir is to extend the theory of combustion that I announced in 1777; to show that Stahl's phlogiston is imaginary and its existence in the metals, sulphur, phosphorus, and all combustible bodies, a baseless supposition, and that all the facts of combustion and calcination

are explained in a much simpler and much easier way without phlogiston than with it. I do not expect that my ideas will be adopted at once . . . I see with much satisfaction that young men, who are beginning to study the science without prejudice . . . who bring fresh minds to bear on chemical facts, no longer believe in phlogiston in the sense that Stahl gave to it and consider the whole of this doctrine as a scaffolding that is more of a hindrance than a help for extending the fabric of chemical science.

As he may have anticipated, Lavoisier's attack on phlogiston brought a firestorm of controversy. Priestley still resented Lavoisier's name for the gas that he had discovered and he flatly rejected the oxygen theory. A colleague of Priestley's, Irish chemist Richard Kirwan, published an essay supporting phlogiston. One of his proofs was that metals release combustible gases when put into acid. The actual substance released is hydrogen, but Kirwan tended to see hydrogen and phlogiston as interchangeable entities. He stated that Lavoisier's antiphlogiston theory was "recommendable in its simplicity," but thought it was "more arbitrary in its application . . . than Stahl's phlogiston theory."

When the Lavoisiers learned about Kirwan's essay, Marie translated it into French. In 1788, they published the translation along with essays by French chemists who had accepted the death of the phlogiston theory.

While professionally Lavoisier fought to disprove the phlogiston theory, at home he had support from his wife, Marie, who continued to be an excellent assistant for her husband. They never had children, but the marriage appears to have been happy. A friend, Jean-Francois Ducis wrote about them:

"Wife and cousin at the same time
Certain to love and to please
For Lavoisier, in thrall to your rule
You fill the two roles
Secretary and muse."

Lavoisier had introduced Marie to chemistry and she later took lessons in the subject from Jean-Baptiste Bacquet. She often served as assistant in the laboratory and took notes on experiments. She learned Latin and English, so she could translate articles and even books into French for Lavoisier to read.

Entertaining guests in their home at the Arsenal was also among Marie's duties and pleasures. Scientists and politicians were always welcome. Benjamin Franklin was a frequent visitor during his stay in Paris.

She also took drawing lessons from a famous French painter, Jacques-Louis David. With his help, she learned to sketch scenes and equipment that were often used to illustrate Lavoisier's memoirs and books.

Marie painted a portrait of Benjamin Franklin and presented it to him as a gift in 1788, to replace a portrait of Franklin that had disappeared during the British occupation of Philadelphia in the American Revolution. Franklin wrote to Marie:

I have a long time been disabled from writing to my dear Friend, by a severe Fit of the Gout, or I should sooner have return'd my Thanks for her very kind Present of the Portrait, which she has herself done me the honour to make of me. It is allow'd by those who have seen it to have great merit as a Picture in every Respect; but what particularly endears it to me, is the Hand that drew it. Our English Enemies when they were in Possession of

Benjamin Franklin

this City and of my House, made a Prisoner of my Portrait, and carried it off with them, leaving that of its Companion, my wife, by itself, a kind of Widow. You have replaced the Husband; and the Lady seems to smile, as well pleased.

In addition to being Marie's teacher, Jacques-Louis David painted a portrait of Antoine and Marie Lavoisier. In the portrait, Antoine is seated at a table with a pen in his hand, as if making notes on a scientific document. Marie leans on his shoulder and he gazes up at her. On the table and at their feet are other scientific instruments; a flask, a hydrometer, a gasometer, and part of the pneumatic trough

Jacques-Louis David

used to collect gases. Behind Marie, her artist's sketching easel stands.

The only unbalance known in the marriage is an affair in which Marie was involved with Pierre-Samuel DuPont. As far as historians know, Antoine was never aware of the affair.

David's portrait of Lavoisier and Marie.

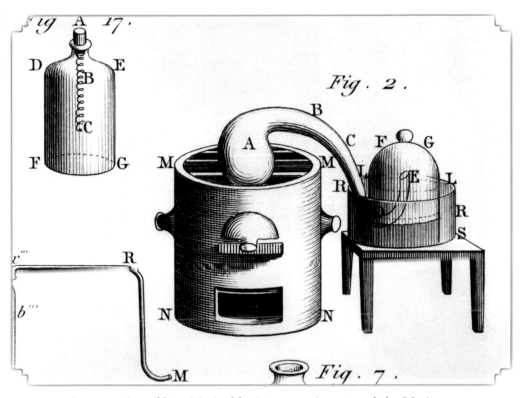

An engraving of Lavoisier's phlogiston experiment, made by Marie.
(Courtesy of The London Art Archive/Alamy)

Some biographers believe that it happened after Lavoisier's death. It became known after Marie's death through letters written by DuPont.

In 1787, Antoine Lavoisier prepared to write and publish his opus, a book entitled *Elements of Chemistry*. The book brought together all of Lavoisier's theories and learning, presenting a clear, unified view of chemistry, and definitively refuted the phlogiston theory. For the book, Marie drew and engraved thirteen plates that were used as illustrations.

When Lavoisier sent his book, *Elements of Chemistry,* to Benjamin Franklin in 1789, he wrote, "I believe, and a

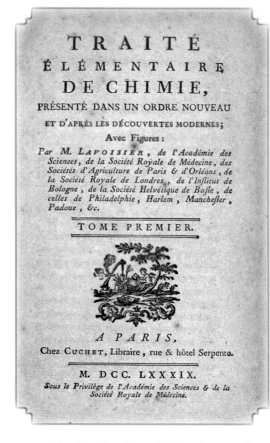

TRAITÉ
ÉLÉMENTAIRE
DE CHIMIE,

PRÉSENTÉ DANS UN ORDRE NOUVEAU
ET D'APRÈS LES DÉCOUVERTES MODERNES;
Avec Figures :

Par M. LAVOISIER, de l'Académie des
Sciences, de la Société Royale de Médecine, des
Sociétés d'Agriculture de Paris & d'Orléans, de
la Société Royale de Londres, de l'Institut de
Bologne, de la Société Helvétique de Basle, de
celles de Philadelphie, Harlem, Manchester,
Padoue, &c.

TOME PREMIER.

A PARIS,
Chez CUCHET, Libraire, rue & hôtel Serpente.

M. DCC. LXXXIX.
Sous le Privilège de l'Académie des Sciences & de la
Société Royale de Médecine.

The frontis of Lavoisier's *Elements of Chemistry*.

great number of scholars today agree with me, that the hypothesis advanced by Stahl, [phlogiston] and since modified, is erroneous, that phlogiston in the sense that Stahl understood the word does not exist, and it is principally to develop my ideas on this subject that I undertook the work which I have the honor to send to you."

Both Antoine and Marie were soon to face their biggest battle, as well. That battle in the midst of the French Revolution would determine who lived and who died.

seven

The Language of Chemistry

I n 1787, Lavoisier presented a lecture at the Easter meeting of the Academy of Sciences. The title was "The need to reform and improve chemical nomenclature." He believed it was time for chemistry to have a new language.

Lavoisier believed that the language of chemistry hindered progress. For example, some compounds had more than one name and some had strange names like pompholix, oil of vitriol, butter of arsenic, and flowers of zinc.

These names made it necessary for chemistry students to memorize long lists of names that had no indication of what elements were contained in the chemicals. As early as the sixteenth century, Agricola had objected to the name litharge, meaning lead oxide. The word litharge comes from a Greek word meaning silver and there is no silver in the compound. Other scientists such as Robert Boyle and Isaac Newton had complained about the complexity and confusion of chemical names, but it was up to Lavoisier and other French scientists to propose the changes that are still in use today.

Guyton de Morveau, a French chemist, had proposed a new chemical language in the past and he and Lavoisier, along with two other chemists, began this work. De Morveau was working on a volume called the *Encyclopédie méthodique.* For this project, de Morveau studied work done by French chemists, as

Lavoisier worked with Guyton de Morveau to develop a new chemical language.

well as papers published by scientists of other countries. He was impressed by essays written by Swedish chemist Torbern Bergman. Bergman had proposed changing the language of chemistry to make it more understandable.

In 1786, de Morveau had finally thrown away his acceptance of phlogiston and accepted Lavoisier's theory of combustion. The destruction of phlogiston was the first shots of the chemical revolution and now a new language would bring

Torbern Bergman

the revolution to all scientists. The project had a nationalistic flavor as it would establish the importance of French chemists over their colleagues in Great Britain and Germany.

Lavoisier had struggled with the language used by chemists in his days as a student. He later wrote, "When I first took up the study of chemistry I was surprised by the number of difficulties that surrounded the approach of this subject, this despite the fact that my instructor taught clearly, was well-disposed towards students, and made every effort to help us understand." He felt that chemistry should be as rigorous and logical as mathematics. "I had in addition become familiar with the rigor with which mathematicians reason in their treatises. . . . Everything is tied together, everything is connected."

He had continued his study of chemistry, but found it confusing. "I managed to gain a clear and precise idea of the state that chemistry had arrived at by that time. Yet it was nonetheless true that I had spent four years studying a science that was founded on only a few facts, that this science was composed of absolutely incoherent ideas and unproven suppositions, that it had no method of instruction, and that it was untouched by the logic of science. It was at this point that I realized I would have to begin the study of chemistry all over again."

Lavoisier had not only studied the science of chemistry; he had begun the transformation of the science with his oxygen theory, but now the language needed an overhaul.

Famous for his pursuit of logic, Étienne Bonnet de Condillac published a book on the subject in 1780. He expanded on the theory of language as a tool of reasoning, an analytic method. Lavoisier stated that Condillac showed "how the language of algebra could be translated into common speech and *vice versa;* how the progress of the understanding is the same in both; and how the art of reasoning is the art of analyzing."

Tying language and reason together amplified the need for a new language for chemistry. The confusing language clouded the relationships and connections found in chemistry.

Children learn language from their earliest days by associating a word with their desires and using the word to get what they want. Lavoisier felt that beginning science students are like children and they need a logical language. However, scientists had been exposed to so many false ideas that have been accepted that their foundation was wrong and a new one was needed. "It is therefore not astonishing, that in the early childhood of chemistry, suppositions instead of conclusions were drawn, that those suppositions transmitted from age to age were changed into presumptions, and that these presumptions were adopted and regarded as fundamental truths even by the ablest minds." Lavoisier proposed to change the movement from guesses to false systems by making the language of chemistry as exact as the mathematical expressions known by everyone. $2 + 2 = 4$ is a basic mathematical fact and chemical language should be just as precise.

Lavoisier stressed that changing the language would change how scientists looked at chemistry:

> We shall have three things to distinguish in every physical science: the series of facts that constitute the science; the ideas that call the facts to mind; and the words that express them. The word should give birth to the idea; the idea should depict the fact; they are three impressions of one and the same seal; and as it is the words that preserve and transmit the ideas, it follows that the science can never be brought to perfection, if the language be not first perfected, and that however true the facts may be and however correct the ideas to which they give

rise, they will still transmit only false impressions, if there are no exact expressions to convey them.

The confusing language of chemistry was a holdover from the time of the alchemists. These men and women had purposely shrouded their language and their experiments in mystery. Even though they developed many valuable chemical facts, their search for the transmutation of lead into gold or the Philosopher's Stone which would cure all illness forced them to keep their experiments hidden from others. So as Lavoisier wrote, "They made use of an enigmatical language."

Lavoisier and de Morveau presented papers to the Academy covering this new language and in 1787, they published a book *Method of Chemical Nomenclature.* In 1789, when Lavoisier published his most famous book, *Elements of Chemistry,* it

A painting of medieval alchemists. Their work was shrouded in mystery.

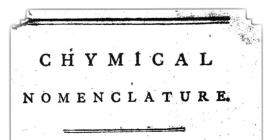

C H Y M I C A L

N O M E N C L A T U R E.

A M E M O I R.

ON THE NECESSITY OF REFORMING AND BRINGING
TO PERFECTION THE NOMENCLATURE OF CHY-
MISTRY; READ TO THE PUBLIC ASSEMBLY OF
THE ROYAL ACADEMY OF SCIENCES IN PARIS,
ON THE 18th OF APRIL, 1787.

By Mr. L A V O I S I E R.

T HE work which we lay before the Academy
has been undertaken in common by Mr. de Morveau, Mr. Bertholet, Mr. de Fourcroy, and by
me: it is the refult of a great number of conful-
tations, in which we have been affifted by the learning and advice of fome geometricians of the Academy, and of feveral chymifts.

A long time before the modern difcoveries had
given a new appearance to the fcience in general,
chymifts perceived the neceffity of giving the no-
B   menclature

Lavoisier published *Method of
Chemical Nomenclature in 1789,*
describing his new chemical
language.

combined the new lan-
guage and the experi-
ments that supported it
into a system to teach and
understand chemistry.

Before naming chem-
icals, Lavoisier and the
other scientists had to
decide what the basic
elements were. In the
second paper read to the
Academy by Guyton de
Morveau in May 1787,
they chose to fall back on
the definition proposed
by Robert Boyle in 1661.
"We shall content our-
selves here with regard-
ing as simple all the substances that we cannot
decompose, all that we
obtain in the last resort
by chemical analysis."

In *Elements of Chemistry,*
Lavoisier listed the elements
known at the time. The heading of the "Table of Simple
Substances" says "Simple substances belonging to all the
kingdoms of nature, which may be considered as the elements of bodies." He lists what he calls the elements with the
new name on the left and the old names on the right. Some
are easily recognized by us today: oxygen, hydrogen, sulfur,

phosphorus. Light and caloric are known to us as energy, but in his day Lavoisier still saw them as real substances.

Part of his list includes substances such as lime and magnesia. Lavoisier realized that these might not be elements, but compounds made up of elements. Other substances that he knew were compounds, he left out. He wrote, "The fixed alkalies, potash, and soda, are omitted in the foregoing Table, because they are evidently compound substances, though we are ignorant as yet what are the elements they are composed of." He was right. In the early 1800s, English chemist Humphrey Davy discovered the elements of potassium and sodium. They are the metallic parts of the compounds of potash and soda.

Once the list of known elements was determined, Lavoisier and de Morveau were ready to tackle new names for compounds. The new language proposed by Lavoisier and the other chemists was similar to the pattern for the living world, created by Swedish biologist Carl Linnaeus in the 1700s. In his system, every living thing has two Latin names because Latin was the language known by all scientists regardless of their nationality. The first name in Linnaeus' system is the genus, similar to a family name. The second is the exact species. For example, *canis familaris,* is the scientific name for dogs. The first name, *canis,* refers to all dogs, wolves, coyotes. *Familaris* describes the animal that has been domesticated by man.

Lavoisier had studied the naming system of Linnaeus as a student at College Mazarin. He felt it would be a good system to adapt for the new chemical language. "It is now time to rid chemistry of every kind of impediment that delays its advance; to introduce into it a true spirit of analysis; and we

Sir Humphry Davy

have sufficiently demonstrated that it is by the perfecting of its language that this reform must be brought to pass."

In the system set up by Lavoisier and the others, the chemical family name comes second and the specific name comes first. Lavoisier thought that all acids contained oxygen, so in the class of acids, the compounds were composed of oxygen

Swedish biologist Carl Linnaeus created a pattern for the living world.

and another element that varied. The family name is acid and the other name tells what other element the compound contains. For example, nitric acid contains nitrogen and oxygen. Although many acids do contain oxygen, today acids are defined as producing hydronium ions (OH^-) in water.

Some acids have the same specific element, but have different amounts of oxygen. Lavoisier added suffixes to the names to distinguish these. So nitric acid has more oxygen than nitrous acid. Today scientists know that the formula for nitric acid is HNO_3 and nitrous acid is HNO_2. In a similar way, Lavoisier named sulfuric acid and sulfurous acid and phosphoric and phosphorous acid.

The suffix –ide means that a compound only has two elements. Examples are zinc oxide, a compound of zinc and oxygen and mercury sulfide, a combination of mercury and sulfur.

Salts are a class of compounds that come from the reactions of acids and bases. For their names, Lavoisier introduced the suffixes, -ate and -ite. The –ate suffix means more oxygen in the compound. For example, sodium nitrate contains more oxygen than sodium nitrite.

In *Elements of Chemistry,* Lavoisier asserted that the new names would be more useful to all chemists and easier for the beginning chemist to understand. "The names, besides, which were formerly employed, such as *powder of algaroth, salt of alembroth, pompolix, phagadenic water, turbith mineral, colcothar,* and many others, were neither less barbarous nor less uncommon. It required a great deal of practice, and no small degree of memory, to recollect the substances to which they were applied, much more to recollect the genus of combination to which they belonged." With the new names, chemists could immediately determine what elements are in a compound. They also could guess how a compound would react based on the reaction of similar compounds.

The new nomenclature was immediately controversial, just as Lavoisier's attack on phlogiston had been. Priestley and Cavendish both rejected the new names. Thomas Jefferson thought it would fail. He said, "it is premature, insufficient, and false . . . and upon the whole I think the new nomenclature will be rejected after doing more harm than good." Other scientists, though, embraced the clarity of the names and the system.

Tucked inside the book, *Elements of Chemistry,* was another paragraph that would have consequences down through the years:

> We must lay it down as an incontestable axiom, that, in all the operations of art and nature, nothing is created; an equal quantity of matter exists both before and after the experiment; the quality and quantity of the elements remain precisely the same; and nothing takes place beyond changes and modifications in the combination of these elements. Upon this principle the whole art of performing chemical experiments depends: We must always suppose an exact equality between the elements of the body examined and those of the products of its analysis.

This is a statement of The Law of Conservation of Mass. By assuming that the same components existed in the chemicals before and after reaction, Lavoisier completely annihilated the ancient idea of transmutation; changing one element into another. Close ideas of this law had been stated by other scientists and Lavoisier had applied it in his experiments for his whole career. However, the plain statement of it in this setting of the new names led chemists to think of reactions in terms of each element involved. Lavoisier again showed his trust in equilibrium, seen through his use of analytical balances in the laboratory. In a beginning chemistry class today, students learn to balance chemical equations, matching the number of atoms of the elements on both sides.

Lavoisier sent two copies of *Elements of Chemistry* to his friend, Benjamin Franklin, who was back in America. One copy was for the American Philosophical Society, of which Lavoisier was a foreign member. Lavoisier presented the other copy to Franklin himself, along with a letter talking about the old theory of phlogiston and the new nomenclature. He knew that Franklin's support would be critical to the acceptance of the naming system.

Lavoisier sent a copy of *Elements of Chemistry* to the American Philosophical Society in Philadelphia, which meets in this building. *(Library of Congress)*

"It seems to me that to present chemistry in this form is to render the study infinitely easier than it has been. Young men whose minds are not preoccupied with any other system seize it with avidity, but old chemists still reject it, and most of these have even more trouble in comprehending and understanding it than those who have not yet made any study of chemistry."

Lavoisier went on to ask for Franklin's support. He wrote, "This then is the revolution which has occurred in an important branch of human knowledge since your departure from Europe; I look upon this revolution as well advanced and it will be complete if you will stand with us."

Lavoisier also sent a copy to Joseph Black, the discoverer of carbon dioxide. "You will find in it some of the ideas of which you sowed the first seed; if you will be good enough to give a little time to reading it, you will find in it the development of a new doctrine which I believe to be simpler than the phlogiston theory and more in agreement with the facts. It is, however, only with trepidation that I submit it to the foremost of my judges, whose approbation I seek above that of all others."

Gradually, scientists from around the world accepted both Lavoisier's new chemical names and his theory of oxygen. Two years later, in a letter to another scientist, Lavoisier wrote, "I take enormous pleasure in seeing that you have adopted the principles I was the first to announce. . . . Letters announcing new converts are arriving from all quarters and I see that the only people who cling to the phlogiston doctrine are those who are too old to take up new investigations or who cannot bend their imaginations to a new way of seeing things. All the young adopt the new theory, which tells me that the revolution in chemistry is over."

As a good French subject, Lavoisier also sent copies of his new book to the king and queen. However, events in the political world were soon to finish the monarchial system of France and while Lavoisier's chemical revolution was over and victorious, the political one was just about to begin.

eight

The French Revolution

L ouis XIV, known as the Sun King, ruled France for seventy-two years. He believed that he was king by a God-given right, and his reign brought an increasingly heavy tax burden and oppression for the people. His son, Louis XV, succeeded him. He was a weak king and his entry into wars cost France large sums of money. Because of the Seven Years War, France lost valuable colonies in Canada and India to Great Britain.

King Louis XVI began his reign in 1774, just as Lavoisier was doing some of his most important research. He was a kinder man than his grandfather, but came across as harsh and stern. He was loved by the people at the beginning of his reign; however, the people of France hated his queen, Marie Antoinette. She was the daughter of the Empress Maria-Theresa of Austria, a country that historically was an enemy of France. Many people called Marie Antoinette "the

King Louis XVI and his queen, Marie Antoinette

Austrian woman." Her lifestyle did nothing to endear her to the people. She spent large sums of money on dresses, gambling, elaborate theatrical productions, and houses. The palace at Versailles was lavishly decorated. One room, the Hall of Mirrors, has gold trimmed mirrors on each wall and exquisite sculptures and paintings.

The king and queen were attended by a lavish court of nobility. Money flowed out of the Treasury to support the court, but little was coming in. The king was broke and needed to find a new way to continue his and his queen's extravagant lifestyle.

Under the ancient régime, the people of France were divided into Three Estates. The First Estate was the clergy and they paid no taxes. This group included the local parish priests up to the archbishops and abbots and together they owned as much as 10 percent of the land of France. The Second Estate was nobility who also paid no taxes. Some of the nobles still worked the land themselves as modest farmers and held to their traditional values of service to the king and their workers. Others lived in a style as lavish as the king. The Third Estate comprised everyone else from poor people to businessmen and lawyers. This group struggled to pay increasing taxes as the country spent more and more money. Historically, these three groups formed the Estates General to govern the country. However, the group had not met for 175 years.

In 1788, King Louis XVI called the Estates General to meet. He hoped they would be able to help him find more wealth. The meeting was set for May 1789.

Lavoisier, as most of the people of France, welcomed the upcoming meeting of the Estates General. Lavoisier's father had bought him a noble title as a wedding gift, but he had not used it. But now as a noble, he traveled to Blois to elect the representatives for that region that would attend the Estates General for the Third Estate. He was appointed secretary and given the duty to write the instructions for the two chosen.

As part of the Provincial Assembly in Orléans, Lavoisier had already considered the type of government that was best for all. He wrote a pamphlet based on his beliefs. "Let us speak frankly; legislative power does not reside in the King alone, but in the concurrence of his will and that of the nation."

In the instructions for the representatives to the Estates General, Lavoisier recommended that they advocate for basic human rights; freedom of the individual, freedom of the press, and reform of the legal system and the taxation system. He also proposed reforms including a new constitution, a national system of education, and regularly scheduled meetings of the Estates General. "We shall, therefore, not take as our guide what our fathers did, for they were wrong; we shall not travel along the road of ancient abuses; the time of enlightenment has come and we must now speak the language of reason and claim those human rights that are inalienable." Lavoisier was seeking for balance in the government of France, just as he did in all areas of life. But this balance would come only at the cost of blood.

Lavoisier was chosen as backup representative in case one of the two could not attend the meeting. Back in Paris, he was active in the '89 Club, an organization that stressed the value of a new constitution.

The Estates General representatives gathered at the king's palace at Versailles on May 5, 1789. The First Estate had 308 representatives; the Second Estate, 285; the Third Estate, 681.

One of the first items of business was the system of voting. The king realized that most of the people of France belonged to the Third Estate, so he had allowed twice as many representatives in that group. However, now he decided that each group would have only one vote, giving the First and Second Estate the power to cancel any vote of the Third Estate.

This ruling inflamed the Third Estate; they already paid most of the taxes and now they could not effectively vote against more taxes or to increase their rights. After days of discussion, the Third Estate rebelled. Some of the representatives of the First Estate sided with them and together they formed a new body called the National Assembly. Three days later, the king locked them out of their meeting place. Instead, they met at an indoor tennis court. In what came to be called The Tennis Court Oath, they swore to stay there until they had written a new constitution. After a few more days, the king relented and let them continue meetings.

The winter of 1788-89 had been extremely harsh on the people of France. The harvest had been poor and many people were hungry. Rumors grew that the king planned to send troops to subdue unrest in Paris. Peasants in the country began attacking castles of the nobles, sometimes burning them and killing the owners. In Paris, a new government called a commune was established. People of the city began tearing down the wall built by the Tax Farm under Lavoisier's recommendation.

On July 14, 1789, a mob attacked the fortress called the Bastille. The Bastille had been a political prison, but when the mob forced the surrender of the garrison soldiers, only seven prisoners were there to be released. The commune wanted the Bastille destroyed and Lavoisier agreed to donate some

After the king locked the National Assembly out of their meeting place, they met at an indoor tennis court, and vowed to write a new constitution.

money to pay a contractor to dismantle it. When King Louis XVI was informed about the fall of the Bastille early on July 15, he asked, "Is this a rebellion?" The person bringing the news said, "No, sire, it is a revolution."

On August 6, Lavoisier was at the Bastille trying to arrange the demolition of the fortress when a crisis over gunpowder erupted. A barge load of gunpowder was taken out of the Arsenal to be shipped to traders in Africa. In French, as in English, the words *trader* and *traitor* are easily confused. A rumor spread that the powder was going to traitors to be used against the people.

Lavoisier was trying to have the Bastille dismantled when a revolutionary mob stormed it. *(Courtesy of The London Art Archive/Alamy)*

Officials made the situation worse by issuing conflicting orders. The powder was unloaded and then loaded and unloaded again. Lavoisier was able to calm the situation by pointing out to city leaders what the destination of the powder was. However, his involvement in the affair wrongly led people to believe he was against the Revolution.

In early 1790, Lavoisier thought the Revolution had accomplished the needed reforms and that the time had come for the country to have calmer times. Still he had some qualms about the future. He wrote to Benjamin Franklin, "After telling you about what is happening in chemistry, it would be well to give you news of our Revolution; we look upon it as over, and well and irrevocably completed. . . . We greatly regret your absence from France at this time; you would have been our guide and you would have marked out for us the limits beyond which we ought not to go."

When Lavoisier wrote to Franklin, the king and his family were under house arrest in Paris, but Lavoisier still hoped that France would form a constitutional monarchy, much like Great Britain's.

In the midst of the anarchy of the country, Lavoisier was given an assignment. The National Assembly asked Lavoisier and others to study standard weights and measurements. Measurements, both length and weight, varied among the different parts of France, causing confusion and adversely affecting trade with other countries. The weights and measures proposed by the group would standardize the system for all of France. Eventually, this system became the metric system, used in most of the world today.

The length measurement, the meter, was based on the length of the circle passing through the earth at the poles. The meter was divided into one hundred smaller units known as centimeters.

Lavoisier's part of the project was to establish the standard for a unit of weight called the gram. He and his colleagues needed to determine the weight of a cubic centimeter (based on the new length standard) of water in a vacuum at 0°C (32°F). That weight would become the new standard, the gram.

The National Assembly also asked Lavoisier to determine the annual wealth of France. He had been interested in this topic for years and he began the work with enthusiasm. He felt that with the report, he could show the Assembly the value of tax reform and good monetary policy. He urged the Assembly to set up a bureau to study the country's economy, so that the facts could help the government make informed choices on how money would be spent. "General accounts of this sort, which could be extended to include studies of the population and the balance of trade, would provide a reliable thermometer of public prosperity. Each legislature would immediately be able to see for the nation as a whole the good and bad consequences of the actions of earlier legislatures."

Soon after Lavoisier's report was published, the Assembly began investigating the Tax Farm. The public accused the Tax Farmers, including Lavoisier, of keeping money that belonged to the government. On March 20, 1791, the Assembly disbanded the Tax Farm.

In spite of the investigation into the Tax Farm, Lavoisier was trusted enough to be appointed to the board of the National Treasury. He took the job without pay, stating that his pay as Gunpowder Commissioner was enough. He hoped to be able to continue to live at the Arsenal where his laboratory was.

On October 1, 1791, the Assembly finished another constitution and instigated new elections. The constitution stated that no one who had served in the old assembly could serve in the new one; this meant that many of the new members had little experience in government, and were unable to alleviate the troubled political climate. The British ambassador wrote, "The present constitution has no friends and cannot last." He was right.

Lavoisier tried to work with the new Assembly, but he realized that the times were dangerous. He wrote to an Englishman, "Today the man who aspires to a great position [in government] must be either very ambitious or very crazy."

Men who had hated Lavoisier's preeminence in science stirred up sentiment against him. Jean Paul Marat had aspired

Jean Paul Marat

to membership in the Academy of Science. In 1780, he presented a paper on combustion. Lavoisier reviewed the paper and said it was wrong. Marat never forgave him. In a newspaper article written by Marat in 1791, he inflamed opinion against Lavoisier by listing "Farmer General, Director of the Gunpowder Administration, administrator of the Discount Bank, member of the Academy of Sciences . . ." Since all these organizations were under scrutiny, Marat implied that Lavoisier was partly responsible for the condition of the country.

Jacques-Pierre Brissot resented Lavoisier's reputation as a scientist and especially hated his role in proving mesmerism false. He harangued Lavoisier in a newspaper article, writing "General Farmer and Academician, two titles for the encouragement of despotism, still worse he is the author of the plan to build a wall around Paris. . . . Lavoisier became a chemist; he would have become an alchemist if he had followed only his unquenchable thirst for gold."

Lavoisier ignored these vicious attacks, but the malevolent words stuck in the memories of the people of Paris. Even after Marat was assassinated in his bathtub on July 13, 1793, his words lingered in the memories of many people.

In August 1792, another mob stormed through Paris. Then, in September 1792, another new government was formed establishing France as a republic. This was the end to the monarchy. The new government even changed the calendar to start with Year One on September 22, 1792.

To establish the republic, the new government decided that the king must die. On January 21, 1793, the king was executed by guillotine. His last words were, "I die innocent . . ." The Reign of Terror, in which revolutionaries began seeking

The Death of Marat, a 1793 painting by Jacques-Louis David, a close friend of Marat.

out and executing anyone associated with the old monarchy, began.

The Academy of Science began being attacked. Many felt it was an elitist society that needed to be dissolved. Lavoisier had been appointed treasurer of the Academy in December 1791, and he defended the Academy as long as he could. He pointed out the valuable work performed by the scientists of the Academy, but on August 8, 1793, all academies were closed.

About the same time, Lavoisier, fearing for his safety, resigned from his position as Gunpowder Commissioner. He and Marie moved out of the Arsenal and into a house on Boulevard de la Madeline.

Meanwhile, executions continued daily. On October 16, 1793, Marie Antoinette, the hated queen, was killed.

Antoine Dupin, the leader of the investigation into the finances of the Tax Farm claimed to have found evidence of crimes committed by the Tax Farm. On November 24, orders went out to arrest all the former Tax Farmers.

Lavoisier felt he could defend himself and the other Tax Farmers if given a little time to prepare. For four days, he hid in the building once occupied by the Academy of Sciences. He wrote letters to influential people. In one letter to the Committee on Public Education, he spoke of himself in third person, writing "Lavoisier, of the former Academy of Sciences, quit the General Farm about three years ago . . . It is public knowledge that he never involved himself in the general affairs of the farm . . . and moreover the works he published attest that he has always principally occupied himself with the sciences."

Lavoisier and his father-in-law, both former members of the Tax Farm and lawyers by education, believed that they would be given a fair trial. On November 28, they surrendered at the prison where the other Tax Farmers were being held. Much of the Lavoisiers' property was confiscated on December 17.

The Tax Farmers were expected to examine all their accounts, and use them to show whether or not they were guilty of aiding the monarchy and extorting the French people, but at the prison where they were held, they had no access to the records. On December 25, the Tax Farm offices themselves were converted into a jail and the men moved there so their accounts would be accessible.

Marie visited her husband and father in prison. After one visit, Lavoisier wrote to her:

> You give yourself, my dear, much trouble, much fatigue of body and soul, and I cannot share it. Take care that you do not impair your health, for that would be the worst of evils. My life is advanced, I have enjoyed a happy existence as long as I can remember, and you have always contributed to it by the devotion that you have shown me; I will leave after me pleasant recollections. My task is done, but you have a right to hope for a long life, so do not throw it away. I thought yesterday that you were sad. Why be so, since I am resigned to all and look upon all that I shall not lose as gained? Besides, we are not without hope of being together again and, while waiting, your visits give me some happy moments.

In January 1794, the Tax Farmers finished going through their accounts. The accounts actually showed that the government owed the Tax Farmers money. Lavoisier thought this proved that they had not been in league with the monarchy, but had been cheated by it just as many of the people had. But without a chance at defense, other charges were placed against them. They were accused of extortion through high interest rates, delaying payments to the Treasury, and adulterating the tobacco sold under their administration.

On May 5, Antoine Dupin convinced the National Convention to put the Tax Farmers on trial by the Revolutionary Tribunal. Most trials in this court ended in a death sentence for crimes against the Republic. In spite of the lack of evidence of treasonous crimes, the Tax Farmers were transferred to the jail called the Conciergie to await their trial.

Marie Antoinette appears before the Revolutionary Tribunal. Lavoisier was later judged by the tribunal, as well.

Lavoisier wrote to a cousin: "I have had a fairly long life, above all a very happy one, and I think that I shall be remembered with some regrets and perhaps leave some reputation behind me." He continued that the events probably meant that he would not suffer from old age, but die "in full possession of my faculties." He ends the letter with, "I am writing to you today, because tomorrow perhaps I may no longer be allowed to do so, and because it is a comfort to me in these last moments to think of you and of those dear to me. Do not forget that this letter is for all those who are concerned about me. It is probably the last that I shall write to you."

On May 8, 1794, the men were brought before the chief judge, Jean-Baptiste Coffinhal. They had only fifteen

Those convicted by the Revolutionary Tribunal were executed with a guillotine. *(Courtesy of The Print Collector/Alamy)*

minutes to consult with their defense lawyers, but no one had any doubt about the verdicts anyway. When a member of the Board of Arts and Crafts pointed out Lavoisier's service to the country as a scientist, Coffinhal is said to have replied, "The Republic has no need for scientists."

As expected, the verdict for all the men was guilty, and they were to be executed immediately. They were loaded into the carts for the ride to the guillotine.

Paulze, Lavoisier's father-in-law, was the third to die. Antoine Lavoisier was fourth; he was fifty-one years old. The bodies were buried in a common grave. The next day, a famous mathematician, Joseph Lagrange, said to a friend, "Only a moment to cut off that head and a hundred years may not give us another like it."

nine
Legacy

In a single day, Marie Lavoisier lost her husband and her father to the guillotine. Most of her husband's property was confiscated. Officials required complete inventories of Lavoisier's lab equipment, supplies, and books. Government men had already searched their house several times, but this time they also inventoried the farm at Fréchines. When the officials were finished, they had inventoried more than 13,000 items, both equipment and specimens. Institutions such as École Polytechnique and the Museum of Natural History received some of Lavoisier's lab supplies, such as mercury and mercuric oxide. The institutions also were allowed to have his equipment, books, and mineral specimens. Even furniture such as bookcases, desks, and chairs were taken, as was a piano.

Dupin continued his investigation of the Tax Farmers, harassing their widows. In September 1794, he released a report saying that the Tax Farmers owed the country money and that their heirs had to pay the loss. Marie lost 2,000 livres income with this judgment. She moved in with a former servant.

The Reign of Terror was slowing down. In July, one of its leaders, Maximilien Robespierre, was guillotined, and finally people began to think about what had happened.

As the furor died down, the widows and children of the Tax Farmers petitioned the government to stop the confiscation of their property until a final report of the accounts could be accomplished. On December 10, 1794, the government agreed with the petition.

Maximilien Robespierre

In papers found after Marie's death was a pamphlet apparently written by her entitled "Information by the Widows and Children of the former Farmers-General against the Representative-of-the-People Dupin." In part because of this pamphlet, an inquiry was opened into the report written by Dupin. When evidence was presented to the National Convention, Dupin was arrested and charged with false reporting and theft. He was later released on amnesty.

Marie was able to get her husband's belongings back. In the spring of 1795, a document ordered the restitution of his furniture, papers, books, scientific instruments, and laboratory apparatus back to the "widow of the unjustly condemned Lavoisier." When the accounts of the Tax Farm were finally balanced correctly, it was learned that the country owed 8 million livres to the men and their families. None of the survivors collected this money, though; they wanted nothing else to do with the Tax Farm.

Marie was thirty-seven years old when her husband was executed. Through the years, she had assisted him in scientific pursuits and entertained guests from around the world. By 1805, she was again entertaining guests, many of them scientists. Sir Charles Blagden, the man who told Lavoisier about Cavendish's discovery of hydrogen, visited Marie often. He proposed marriage to her, but she refused.

Another visiting scientist was Sir Benjamin Thompson, Count von Rumford. Thompson was born in Massachusetts, but chose the British side during the American Revolution. He migrated to England and was knighted there. Later he served the government of Bavaria and became Count von Rumford. On October 22, 1805, Marie married Rumford. She insisted

After Lavoisier's death, Marie married Sir Benjamin Thompson, Count of Rumford. *(Courtesy of mediacolor's/Alamy)*

on being called Countess Lavoisier-Rumford. The marriage was not a happy one and in 1809, the couple divorced.

Marie died suddenly on February 10, 1836, forty-two years after Antoine. With the deaths of Antoine and Marie, who had no children, the family line ended.

The rest of Antoine Lavoisier's legacy can be seen by opening any modern day chemistry textbook. In a chapter about combustion, one will find Lavoisier's oxygen theory, stated now much like he did in 1772. The chapter on chemical names will include acids with endings like –ous and –ic and salts with endings of –ite and –ate. The chemical names have two parts just as Lavoisier designed and the system has been flexible enough through the centuries to incorporate elements that Lavoisier had never seen.

His definition of what an element was, and his list of elements prepared chemists to look for others and to determine which ones on the list were compounds rather than elements.

Lavoisier may not have been as innovative an experimenter as men like Priestley and Black, but his talent was seeing the big picture. He realized what the experimental results meant. With his intuitive leaps, he was able to see the common link of combustion, calcination, and respiration. He knew that the link was the gas that was absorbed.

Another legacy is in the names of gases; Lavoisier is responsible for naming oxygen and hydrogen. He also determined that living matter was made up of carbon, hydrogen, phosphorus, and oxygen. His model of respiration was close to the process as it is known today.

Lavoisier was not always careful to give credit to the discoveries of other scientists. Priestley and Cavendish both claimed that he received accolades for their work. However, they might have made the discoveries, but Lavoisier put it together into the oxygen theory.

Lavoisier always paid strict attention to the weights and measurements of the components of his experiments. By meticulous use of the balance, he was able to state the Law of Conservation of Mass. He introduced the idea of a balanced reaction. His insistence on experiments to uncover facts was critical to his success.

The destruction of phlogiston cleared the way for the discoveries and theories of other scientists besides Lavoisier. In the years immediately following his death, scientists such as John Dalton, Joseph Proust, and Amedeo Avogadro advanced Lavoisier's ideas into the Golden Age of Chemistry.

John Dalton

Drawn & Etched by J. Stephenson.

Joseph Proust

Lavoisier also worked to improve the lives of his countrymen through better farming techniques, cleaner hospitals and prisons, purer water, and fairer taxes. He managed to balance science and service to his country.

Lavoisier had instigated and carried through his chemical revolution, and the world of science changed completely because of him. "After 1789 the majority of chemists still worked with the same chemicals, but when they looked at a chemical reaction they did not think of essences and principles swirling around in the flask. Instead, they pictured chemicals combining and recombining with other chemicals."

Amadeo Avogadro

Lavoisier led his chemical revolution in the midst of the French Revolution which cost him his life. However, soon after the Revolution had ended, Lavoisier's reputation was restored. A monument in his honor was erected by the French government about a hundred years after his death.

In 1943, while France was under Nazi occupation, an exhibition of Lavoisier's scientific apparatus was held to mark the

A commemorative coin issued in 1948, remembering Lavoisier.
(Courtesy of The Print Collector/Alamy)

two-hundredth anniversary of his birth. Many of the instruments from his laboratory are on display at the Musée des Arts et Métiers in Paris. In 2003, some of the equipment was used in a reenactment of his experiments. His scientific papers are stored at the Archives of the Academy of Sciences, housed in the building that had been the College Mazarin where Lavoisier attended school.

The two revolutions experienced in France of the 1700s impacted Antoine Laurent Lavoisier's life. The chemical revolution he instigated and encouraged. The French Revolution took his life at a young age. But his memory lives on in his achievements.

A statue of Lavoisier, at the Louvre in Paris

Timeline

1743   Born in Paris on August 26.

1760-

1763   Studies at University of Paris; attends scientific
             lectures; becomes assistant to Jean Ètienne Guettard.

1766   Awarded gold medal from the king for essay
             about street lighting.

1767   Accompanies Guettard on survey of the
             Vosges Mountains.

1768   Admitted into Academy of Sciences; invests in
             Tax Farm.

1771   Marries Marie Paulze on December 16.

1774   Publishes *Physical and Chemical Essays*; appointed
             Gunpowder Commissioner of France.

1777   Gives breathable air the name "oxygen."

1778   Buys farm in Fréchines.

1783   Walls built around Paris at Lavoisier's
             recommendation.

1785   Raises objections to phlogiston theory in paper delivered to the Academy of Sciences.

1787   Works to reform language of chemistry to be more uniform and precise.

1788   Becomes director of Discount Bank.

1789   Publishes *Elements of Chemistry*; attends meeting of Estates General.

1790   Helps develop standard system of weights and measures, the metric system.

1794   Executed by guillotine on May 8, during Reign of Terror.

1795   Lavoisier's property, confiscated by Assembly during Reign of Terror, returned to Marie, with note claiming he was "unjustly condemned."

Sources

CHAPTER ONE: Early Life

p. 14, "that the nature of water . . ." Arthur Donovan, *Antoine Lavoisier: Science, Administration, And Revolution* (Oxford, England: Blackwell Publishers, 1993), 93.

p. 19, "That is one of my secrets . . ." Douglas McKie, Ph.D, *Antoine Lavoisier: Scientist, Economist, Social Reformer* (New York: Henry Schuman, 1952), 20.

p. 20, "I am young and avid . . ." *Eric Weisstein's World of Scientific Biography*, http://www.scienceworld.wolfram. com.

CHAPTER TWO: Early Science

p. 23, "There is a certain sequence . . ." Donovan, *Antoine Lavoisier: Science, Administration, And Revolution,* 84.

p. 24, "this flattering distinction for . . ." McKie, *Antoine Lavoisier: Scientist, Economist, Social Reformer,* 63.

p. 26, "when I get back to Paris . . ." Donovan, *Antoine Lavoisier: Science, Administration, And Revolution,* 87.

p. 27, "I have never seen anything . . ." McKie, *Antoine Lavoisier: Scientist, Economist, Social Reformer,* 68.

p. 27, "Fear about the mines at . . ." Ibid., 66.

p. 28, "Send news as often as . . ." Ibid.

p. 30, "He regrets that you were not . . ." Donovan, *Antoine Lavoisier: Science, Administration, And Revolution,* 41.

p. 31, "Chemists have plenty of ways . . ." Ibid., 88.

p. 31, "the hydrometer can be wonderfully . . ." Ibid.

p. 31, "This part of chemistry is . . ." Ibid., 89.

p. 34, "a fool, . . .a kind of ogre . . ." McKie, *Antoine Lavoisier: Scientist, Economist, Social Reformer,* 92.

CHAPTER THREE: Water, Fire, and Air

p. 35, "Reason must continually be subjected . . ." Donovan,
*Antoine Lavoisier: Science, Administration, And
Revolution,* 71.

p. 41, "out of them, and put . . .back . . ." Madison Smartt
Bell, *Lavoisier in the Year One* (New York: Atlas
Books, 2005), 72.

p. 41, "It is easy to see that this System . . ." Ibid.

p. 41, "It seems constant that air enters . . ." Ibid., 73.

p. 42, "fire enters into the composition . . ." Ibid., 77.

p. 42, "What we will say concerning air . . ." Donovan,
*Antoine Lavoisier: Science, Administration, And
Revolution,* 100.

p. 42-43, "The solution to this problem tends . . ." Bell,
Lavoisier in the Year One, 77.

p. 45, "this increase of weight arises . . ." McKie, *Antoine
Lavoisier: Scientist, Economist, Social Reformer,* 102.

p. 45-46, "large quantity of air was liberated . . ." Ibid.

p. 46, "observed differences so great between . . ." Ibid., 103.

p. 46, "come far short of the number necessary . . ." Ibid., 104.

p. 46-47, "The importance of the end . . ." Ibid.

p. 47, "It obviously results from these experiments . . ."
Bell, *Lavoisier in the Year One,* 90.

p. 47, "above all decisive experiments have . . ." Ibid.

p. 47, "I have even come to the point . . ." Ibid., 91.

p. 48, "I should like to hear . . ." Ibid.

CHAPTER FOUR: The Oxygen Theory

p. 52, "The feeling of it to my lungs . . ." McKie, *Antoine
Lavoisier: Scientist, Economist, Social Reformer,* 118.

p. 55, "Here is the most complete sort . . ." Bell, *Lavoisier
in the Year One,* 107.

p. 55, "I shall henceforward designate dephlogisticated . . ."
Ibid., 108.

p. 56, "Besides, since I am at the point . . ." Frederic
Lawrence Holmes, *Lavoisier and the Chemistry of
Life* (Madison, Wisconsin: The University of Wisconsin
Press, 1985), 106.

p. 59, "Thus the air that we breathe . . ." McKie, *Antoine
Lavoisier: Scientist, Economist, Social Reformer,* 145.

p. 60, "If the animal did not receive . . ." Lisa Yount, *Antoine
Lavoisier: Founder of Modern Chemistry* (Springfield, NJ:
Enslow Publishers, Inc., 1997), 81.

p. 61, "condensed into the dew . . ." McKie, *Antoine
Lavoisier: Scientist, Economist, Social Reformer,* 164.

p. 61, "water in a very pure state." Ibid., 167.

p. 62, "water is not an element . . ." Bell, *Lavoisier in the
Year One,* 117.

CHAPTER FIVE: Government Work

p. 63, "It was for him a day of happiness . . ." McKie, *Antoine
Lavoisier: Scientist, Economist, Social Reformer,* 255.

p. 65, "Happily chemistry provides a reliable . . ." Donovan,
*Antoine Lavoisier: Science, Administration, And
Revolution,* 123.

p. 68, "One can truly say that North America . . ." Ibid., 199.

p. 68, "It is less the loss of a father . . ." McKie, *Antoine
Lavoisier: Scientist, Economist, Social Reformer,* 131.

p. 69, "Lavoisier . . .is a great and good man . . ." Vivian
Grey, *The Chemist Who Lost His Head: The Story of
Antoine Laurent Lavoisier* (New York, Coward-
McCann, Inc., 1982), 57.

p. 70, "Increase in price decreases consumption . . ."
McKie, *Antoine Lavoisier: Scientist, Economist,
Social Reformer,* 221

p. 70, "a fluid whose movements necessarily . . ." Bell,
Lavoisier in the Year One, 17.

p. 73-74, "Nature, left to her own devices . . ." McKie, *Antoine Lavoisier: Scientist, Economist, Social Reformer,* 198.

p. 74, "We discovered we could influence . . ." Walter Isaacson, *Benjamin Franklin: An American Life* (New York, Simon & Schuster, 2003), 427.

p. 74, "Some think it will put an end . . ." Ibid.

CHAPTER SIX: Attack on Phlogiston

p. 79-80, "It is evident that Stahl's . . ." Donovan, *Antoine Lavoisier: Science, Administration, And Revolution,* 103.

p. 80, "matter of fire is a very subtle . . ." Ibid., 151.

p. 82, "All these reflections confirm . . ." McKie, *Antoine Lavoisier: Scientist, Economist, Social Reformer,* 156.

p. 82-83, "It is time to lead chemistry . . ." Ibid., 157.

p. 83, "the molecules of bodies . . ." Donovan, *Antoine Lavoisier: Science, Administration, And Revolution,* 171.

p. 83-84, "My only object in this memoir . . ." McKie, *Antoine Lavoisier: Scientist, Economist, Social Reformer,* 157.

p. 84, "recommendable in its simplicity . . ." Bell, *Lavoisier in the Year One,* 130.

p. 85, "Wife and cousin at the same time . . ." Ibid., 15.

p. 85-86, "I have a long time been disabled . . ." McKie, *Antoine Lavoisier: Scientist, Economist, Social Reformer,* 95.

p. 90, "I believe, and a great number of scholars . . ." Bell, *Lavoisier in the Year One*, 147.

CHAPTER SEVEN: The Language of Chemistry

p. 94, "When I first took up the study . . ." Donovan, *Antoine Lavoisier: Science, Administration, And Revolution,* 46.

p. 94, "I had in addition become familiar . . ." Ibid., 47.

p. 94, "I managed to gain a clear . . ." Ibid.

p. 94-95, "how the language of algebra . . ." McKie, *Antoine*

Lavoisier: Scientist, Economist, Social Reformer, 265.

p. 95, "It is therefore not astonishing . . ." Ibid., 266.

p. 95-96, "we shall have three things . . ." Ibid., 267.

p. 96-97, "They made use of an enigmatical . . ." Ibid., 268.

p. 97, "We shall content ourselves here . . ." Ibid., 269.

p. 98, "The fixed alkalies, potash, and soda . . ." Antoine Lavoisier, *Elements of Chemistry* (New York: Dover Publications, Inc., 1965), 178.

p. 98-99, "It is now time to rid chemistry . . ." McKie, *Antoine Lavoisier: Scientist, Economist, Social Reformer,* 269.

p. 101, "The names, besides, which were formerly employed . . ." Lavoisier, *Elements of Chemistry,* XXXI.

p. 102, "it is premature, insufficient . . ." Bell, *Lavoisier in the Year One,* 138.

p. 102, "We must lay it down . . ." Ibid., 130.

p. 104, "It seems to me that to present chemistry . . ." Bell, *Lavoisier in the Year One,* 148.

p. 104, "This then is the revolution . . ." Ibid., 149.

p. 104, "You will find in it some of the ideas . . ." McKie, *Antoine Lavoisier: Scientist, Economist, Social Reformer,* 306.

p. 104, "I take enormous pleasure . . ." Donovan, *Antoine Lavoisier: Science, Administration, And Revolution,* 185.

CHAPTER EIGHT: The French Revolution

p. 109, "Let us speak frankly . . ." McKie, *Antoine Lavoisier: Scientist, Economist, Social Reformer,* 293.

p. 109, "We shall, therefore, not take . . ." Ibid., 294.

p. 111, "Is this a rebellion?" Phyllis Corzine, *The French Revolution* (San Diego: Lucent Books, 1995) 46.

p. 113, "After telling you about what is happening . . ." McKie, *Antoine Lavoisier: Scientist, Economist, Social Reformer,* 308.

p. 114, "General accounts of this sort . . ." Donovan, *Antoine*

Lavoisier: Science, Administration, And Revolution, 209.

p. 115, "The present constitution has no friends . . ." Ibid., 264.

p. 115, "Today the man who aspires . . ." Yount, *Antoine Lavoisier: Founder of Modern Chemistry,* 86.

p. 116, "Farmer General, Director of the Gunpowder . . ." Bell, *Lavoisier in the Year One,* 161.

p. 116, "General Farmer and Academician, two titles . . ." Ibid.

p. 116, "I die innocent . . ." Phyllis Corzine, *The French Revolution,* 83.

p. 118, "Lavoisier, of the former Academy of Sciences . . ." Bell, *Lavoisier in the Year One,* 176.

p. 119, "You give yourself, my dear . . ." McKie, *Antoine Lavoisier: Scientist, Economist, Social Reformer,* 383.

p. 120, "I have had a fairly long life . . ." Ibid., 400.

p. 120, "I am writing to you today . . ." Ibid.

p. 121, "The Republic has no need . . ." Yount, *Antoine Lavoisier: Founder of Modern Chemistry,* 100.

p. 122, "Only a moment to cut off that head . . ." McKie, *Antoine Lavoisier: Scientist, Economist, Social Reformer,* 407.

CHAPTER NINE: Legacy

p. 125, "widow of the unjustly condemned Lavoisier," McKie, *Antoine Lavoisier: Scientist, Economist, Social Reformer,* 425.

p. 129, "After 1789 the majority of chemists . . ." Hugh W. Salzberg, *From Caveman to Chemist: Circumstances and Achievements* (Washington, DC: American Chemical Society, 1991), 204.

Bibliography

Bell, Madison Smartt. *Lavoisier in The Year One.* New York: W. H. Norton & Company, 2005.

Corzine, Phyllis. *The French Revolution.* San Diego: Lucent Books, 1995.

Donovan, Arthur. *Antoine Lavoisier: Science, Administration, And Revolution.* Oxford, England: Blackwell Publishers, 1993.

Gray, Vivian. *The Chemist Who Lost His Head: The Story of Antoine Laurent Lavoisier.* New York: Coward-McCann, Inc., 1982.

Holmes, Frederic Lawrence. *Lavoisier and the Chemistry of Life.* Madison, Wisconsin: The University of Wisconsin Press, 1985.

Isaacson, Walter. *Benjamin Franklin: An American Life.* New York: Simon & Schuster, 2003.

Lavoisier, Antoine. *Elements of Chemistry.* New York: Dover Publications, Inc., 1965.

McKie, Douglas. *Antoine Lavoisier: Scientist, Economist, Social Reformer.* New York: Henry Schuman, 1952.

Salzberg, Hugh W. *From Caveman to Chemist: Circumstances and Achievements.* Washington, DC: The American Chemical Society, 1991.

Yount, Lisa. *Antoine Lavoisier: Founder of Modern Chemistry.* Springfield, NJ: Enslow Publishers, Inc., 1997.

Web sites

Woodrow Wilson Leadership Program in Chemistry, The Chemical Heritage Foundation
http://www.chemheritage.org/classroom/chemach/forerunners/lavoisier.html
A brief biography of Lavoisier, along with links to pages about other scientist and chemical innovations.

http://moro.imss.fi.it/lavoisier/Index.htm
Visitors to this site will find a virtual museum of the collections of the French chemist scattered throughout the world, including a catalogue of Lavoisier's manuscripts.

University of Virginia
http://cti.itc.virginia.edu/~meg3c/classes/tcc313/200Rprojs/lavoisier2/home.html
University of Virginia page about Lavoisier, with quick, bullet-point explanations of his key theories.

Index